HOW TO HANDLE THE AI REVOLUTION

HOW TO HANDLE THE AI REVOLUTION

BY
DR. LAWRENCE CAPRI

WONDERNOTE PUBLISHING

First Printing. Printed on Acid-free paper and produced and bound
in the United States of America.

FSC, SFI and PEFC Certified
Place of publication Berkeley, California
Cover Design by Keno Mapp

ISBN: 979-8-9918606-0-4
US Copyright Number: TXu 2-466-354
Library of Congress Control Number: 2025902856

For additional information:
www.aimyeverything.com
www.youtube.com/@Dr.LawrenceCapri

*To the dreamers, the explorers,
and the future makers.
To the children stepping into a world filled
with the wonders of AI.
May you see the potential, not the fear.
May you use these tools to uplift, innovate,
and inspire, shaping a world where technology
serves humanity's highest purpose.*

*The future will look different, but it will be
brighter because of you.*

TABLE OF CONTENTS

PREPARATION   24
Building a Futureproofing Mindset

ARTISTS AND AI   31
Collaborating with Creativity

AI AND JOBS

Navigating the Future of Work

AI AND FAMILY DYNAMICS

More Family Time

AI AND HEALTHCARE
Revolutionizing Medicine for Families

AI AND EDUCATION
Preparing the Next Generation

AM I TOO OLD?   86
Navigating the AI Revolution at Any Age

FAMILY LIFE AND AI   93
A New Kind of Connection

AI AND THE FUTURE OF ART AND CREATIVITY   96
The Pope Dj

PREPARING FOR THE AI REVOLUTION   103
Skills and Adaptability in a Rapidly Changing World

AI ETHICS AND RESPONSIBILITY   107
With Great Power Comes Great Responsibility

PREPARING FOR THE AI-POWERED FUTURE   111
Change On The Horizon

AI ACCEPTANCE   144

*Navigating the Emotional and Practical Challenges
of AI Adoption*

YOUR PERSONAL ASSISTANT   149

Embracing AI as a Partner, Not a Threat

A NEW HOPE   155
How AI Can Improve the Future for Humanity

RESOURCES FOR AI REVOLUTION   161
Get Going! You're Already Late

PREFACE

You're standing at the dawn of a new era—an era defined by Artificial Intelligence. The world is changing faster than we ever imagined, and here you are, ready to step into the future. AI is no longer just a buzzword; it's the driving force behind innovations that are reshaping every aspect of our lives. But this isn't just about technology—it's about *you*.

The AI revolution isn't something happening *to* us, it's happening *with* us. It's an invitation to rethink what's possible, to push the boundaries of your creativity, your work, and your personal growth. Whether you're an artist, a professional, a parent, or simply someone curious about the future, this book will help you navigate the opportunities and challenges of AI with confidence and optimism.

Perhaps you're wondering: How do I begin? How can I harness this power for good? Will AI enhance my life or complicate it? These are the very questions that inspired this book. Together, we'll explore the world of AI, not as distant observers, but as active participants, fully prepared to embrace its transformative potential.

Artificial Intelligence will challenge you, inspire you, and expand your understanding of what's possible. It will revolutionize how we work, learn, and connect with others. It's not about machines overtaking humanity—it's about enhancing our collective potential. The possibilities are endless, and the most important tool you have is an open mind.

So, let's embark on this journey together. Whether you're new to AI or already familiar with its capabilities, you'll find insights and practical advice here to help you navigate the future with purpose and excitement. This is more than just a guide—it's a roadmap to thriving in a world where AI and human creativity come together to create something truly extraordinary.

The revolution has already begun. The question is: *Are you ready?*

Let's dive in.

EMBRACING THE DAWN OF THE AI REVOLUTION

We stand at the threshold of a new era—an age where Artificial Intelligence is no longer science fiction, but a transformative force shaping every aspect of our world. It's as if humanity has unlocked a new level of possibility, one where innovation is limitless, and the future teems with opportunities for growth and evolution. But with this new power comes uncertainty, curiosity, and even fear. How do we make sense of this sweeping change? How do we ensure that this revolution benefits not just the tech-savvy, but everyone?

This book is a guide through that journey, a beacon to help you not only understand AI but to thrive in its wake. You may be wondering, "Is AI going to replace me, or can it help me become a better version of myself?" The answer lies not in fearing this new technology but in learning how to harness it. AI isn't here to control us—it's here to collaborate with us, enhancing our creativity, our intelligence, and our ability to solve problems we once thought insurmountable.

You don't need to be an expert in coding or have a background in technology to benefit from what lies ahead. Whether you're an artist, a parent, a student, or someone simply trying to keep pace with change, this revolution is for you. The tools, insights, and innovations we'll explore in this book are designed to empower you, regardless of your age, your profession, or your starting point.

What makes this moment in history so unique is not just the technology itself, but the shift in mindset that it invites. We are no longer passive consumers of information or technology—we are participants in an evolving relationship with AI, one that can unlock doors to success, creativity, and new ways of living. It's time to take the leap, to not only accept but embrace this new reality. Together, we'll explore how to navigate this journey, how to evolve, and how to seize the opportunities that AI offers us all.

This is not just a book about technology; it's a book about hope. It's about the extraordinary possibilities that await when we allow ourselves to adapt and grow with this revolution. As you turn these pages, you're not just learning about AI—you're learning how to use it to become the person you've always wanted to be. Welcome to the future.

Let's get started.

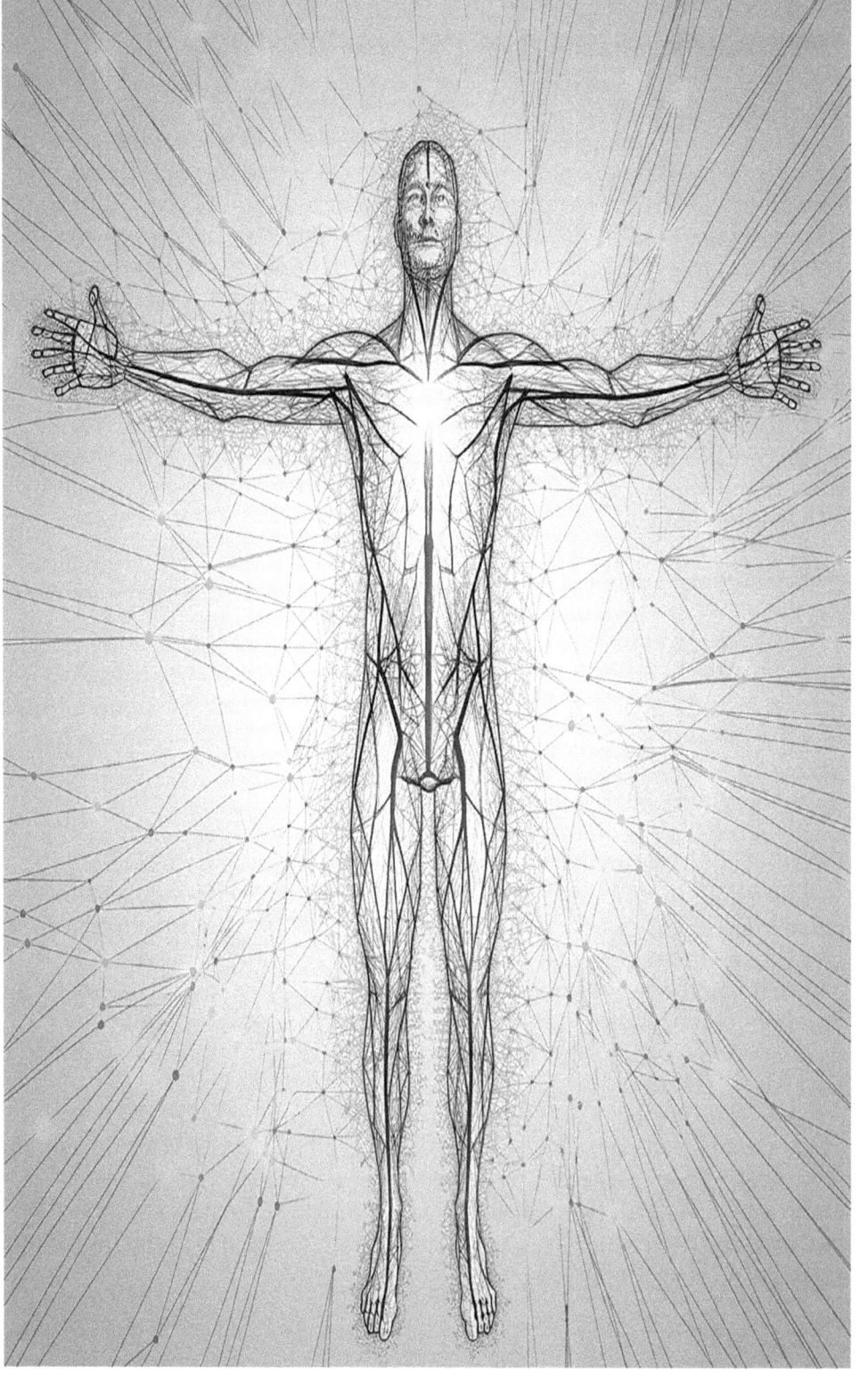

Chapter 1

WHAT'S HAPPENING / WHAT IS AI?

Introduction:
We Are Living in the Future

I magine waking up in the morning, and before you even get out of bed, you ask your virtual assistant to read you the news.Your coffee maker has already started brewing your morning cup because it knows you always drink coffee at 7 a.m. sharp. When you hop into your car, it maps out the best route to avoid traffic based on real-time data. By the time you sit down at your desk, you're ready to tackle your day, armed with insights that AI-powered systems provided you overnight.

For many of us, this isn't the future—it's our present reality. And yet, despite how embedded AI is in our lives, it still feels like something out of science fiction. The reality is, we're living in a world where **artificial intelligence** is making decisions for us, shaping our experiences, and changing the way we interact with the world, often without us even realizing it.

But what exactly is AI?
How does it work?
And more importantly, how is it going to impact you?

WHAT EXACTLY IS AI?

Imagine for a moment that you have a friend who is incredibly good at solving puzzles. You give them the same type of puzzle over and over, and each time, they solve it faster and more accurately than before. This is kind of what Artificial Intelligence (AI) is—a system designed to learn and improve over time by solving specific tasks.

In the simplest terms, *AI* refers to any machine or software that mimics human intelligence. It's about making computers think like us—or at least, think in ways that can help us get things done faster and more efficiently.

Now, while AI may seem like a mystical force that only tech geniuses understand, it's actually much simpler when broken down. There are two main types of AI, and understanding them will make everything clearer:

1. Narrow AI: This is what's all around us right now. Narrow AI is like a specialist. It can do one thing really well, but that's all it can do. Think of Siri on your iPhone. It's amazing at answering questions like "What's the weather?" or "Play my favorite song, " but ask it to fix a car or write a novel, and it's utterly useless. Narrow AI is highly focused on a specific task, whether it's facial recognition, voice commands, or playing chess.

2. General AI: This is where things get futuristic. General AI is the dream of creating a machine that can think, reason, and learn across many areas—just like a human. It would be able to solve any problem, adapt to any situation, and even potentially hold a conversation that feels indistinguishable from talking to a person. The good news (or bad, depending on your perspective) is that this level of AI doesn't exist yet. For now, we're dealing with systems designed for narrow, specific tasks.

A PERSONAL CONNECTION TO AI

Let me share a story with you: A friend of mine, Sarah, was completely uninterested in technology. For her, AI sounded like something straight out of a sci-fi movie—cool, but distant from her everyday reality. Then something unexpected happened: Sarah's father was diagnosed with Alzheimer's.

Sarah had heard about AI being used in healthcare, so she did some research and found an AI-powered app that could help track his symptoms and offer suggestions based on patterns it noticed. It wasn't some futuristic robot doctor—just a simple tool on her phone that made her dad's care more manageable. Over time, the app started to predict what times of day he would need more assistance or when he was likely to become confused, helping Sarah plan her day around his needs.

For Sarah, AI went from something abstract to something deeply personal and practical. And that's the reality we're in. AI is no longer just for scientists or tech enthusiasts—it's here to help each and every one of us in ways we might not have imagined.

DEBUNKING THE MYTHS: IT'S NOT AS SCARY AS YOU THINK

There's a lot of fear surrounding AI, and it's no surprise—Hollywood has given us plenty of reasons to be skeptical, from "The Terminator" to "Ex Machina". But let's take a moment to break down some of the common misconceptions about AI.

1. AI is going to take over the world!

   AI isn't a sentient, all-knowing being that's going to take control of humanity. Right now, AI is a tool—like your smartphone or your computer. It follows instructions, it solves problems, but it doesn't "want" anything. It's not conscious, it's not plotting a takeover. The systems we have today are good at specific tasks, but they can't make decisions beyond what they're programmed to do.

2. AI is going to steal all our jobs!

This is one of the most common fears, but the reality is more complex. Yes, AI will change the job market—just like the Industrial Revolution did. But that doesn't mean it's all bad news. While some jobs will be replaced, AI will also create entirely new roles and industries. Think about how the internet changed things. Jobs like "social media manager" didn't exist 20 years ago. AI will lead to similar shifts, creating opportunities for those who are prepared.

3. AI is too complicated for me.

Not true! You don't need to be a tech wizard to use AI. In fact, you're probably already using it. Do you use Google Maps? That's AI helping you navigate traffic. Do you stream shows on Netflix? That's AI recommending content based on your tastes. AI is meant to make life easier, not more complicated. You don't have to understand how it works to benefit from it—just like you don't need to know how an engine works to drive a car.

THE JOURNEY OF AI: FROM DREAM TO REALITY

Let's take a moment to step back in time. The concept of AI has been around for decades, and what we're seeing today is the result of years of trial, error, and innovation.

The first big leap came in 1956, when scientists at Dartmouth College held the first official AI conference. They believed that, within a few years, machines would be able to mimic human intelligence. That turned out to be overly optimistic, as progress was much slower than anticipated. In fact, AI went through several "winters," where interest and funding dried up because the technology wasn't delivering on its promises.

Fast forward to the early 2000s, and things started to change. With the rise of *big data*—the massive amounts of information being generated daily by the internet—and *increased computing power*,

AI had the fuel it needed to evolve. Suddenly, machines weren't just following pre-set rules anymore—they were learning. This breakthrough in **machine learning** meant that AI could improve over time, becoming more accurate and efficient the more data it processed.

Now, we're in the middle of an AI renaissance. Technologies like *deep learning* and *neural networks* are allowing AI to do things that seemed impossible just a few decades ago—like understanding natural language, recognizing faces, and even generating art.

AI IN ACTION: HOW IT'S ALREADY CHANGING YOUR LIFE

You might not realize it, but AI is already a big part of your daily routine. Let's take a look at some examples you probably encounter without even thinking about it:

Your Smartphone: Every time you use your phone's facial recognition to unlock it, AI is at work. It's analyzing your features to make sure it's you and not someone else trying to gain access.

Social Media: Platforms like Instagram and Facebook use AI to curate your feed. Based on what you like, comment on, or spend time looking at, AI algorithms determine what content you're most likely to engage with, creating a personalized experience.

Streaming Services: Netflix and Spotify use AI to recommend shows, movies, and music based on your preferences. The more you watch or listen, the better the recommendations get.

Customer Service: Ever had a chat with an online customer service bot? That's AI using natural language processing to try and help you with your issue before a human steps in.

WHY NOW? THE AI EXPLOSION

So, why is AI suddenly such a hot topic? There are a few reasons for this:

Data, Data, Data: We live in the age of big data. Every time you send a text, post a photo, or search for something online, you're generating data. AI thrives on data—it's how it learns and improves. With so much data now available, AI has more "fuel" than ever before.

Powerful Computers: The kind of AI we have today requires a lot of computing power. Think of it as a super-fast brain that can process enormous amounts of information in seconds. Thanks to advances in hardware, we now have the technology to support AI at a large scale.

Improved Algorithms: In recent years, AI algorithms have gotten better at learning. These new algorithms allow machines to understand complex patterns, which is why AI can now do things like recognize speech, diagnose diseases, and even generate art.

THE FUTURE OF AI: WHAT COMES NEXT?

The AI of today is just the beginning. Imagine a world where:

AI can help solve global issues, like climate change by optimizing energy use or helping predict natural disasters before they happen.

AI becomes your creative partner, assisting you in making music, writing stories, or designing art.

AI empowers people with disabilities to live more independently, through tools that assist with communication, mobility, and even sensory enhancement.

We're at the start of something massive, and the opportunities are endless.

CONCLUSION: UNDERSTANDING THE REVOLUTION

The AI revolution is already underway, and it's changing the way we live, work, and think. The key isn't to fear it, but to understand it. The more we know about AI, the better prepared we are to harness its potential for good.

In the next chapter, we'll explore how AI impacts *you* on a personal level and how it can be a tool to help you grow, adapt, and succeed in this ever-changing world.

Chapter 2

WHY DO I NEED IT? / HOW CAN IT HELP?

The AI Revolution
The Power of Embracing AI

Let's take a moment to pause and reflect: Do you remember when the internet was first becoming popular? There were skeptics, right? People who thought it was just a fad. But look where we are now—the internet is not just an add-on to life; it's a vital part of how we live, work, and connect.

AI is on a similar path. At first, it might feel foreign or even unnecessary. But just like the internet, it's quickly becoming something you can't imagine living without. The truth is, AI is already helping you, even if you don't realize it. From managing everyday tasks to opening up entirely new ways of thinking, AI is shaping the future—and embracing it will empower you to stay ahead.

WHY DO I NEED AI?

Let's start with the question that's likely on your mind: Why do you need AI? Isn't it just something tech companies use to power their cool gadgets and services? Sure, companies like Google, Amazon, and Apple are pushing AI innovations. But more importantly, AI is rapidly transforming every corner of our personal and professional lives.

Whether you're an artist, a business owner, a teacher, or just someone trying to manage day-to-day tasks,

AI can help you work smarter, not harder. *The point of AI is to enhance human capability, not replace it.

A PERSONAL STORY: "I NEVER THOUGHT I'D USE AI"

Meet *David*, a small business owner. When he first heard about AI, he thought it was for big tech companies—not for him. His company sells handmade candles, and he didn't see how AI would play a role in that. But when the pandemic hit, his business took a huge blow. Foot traffic disappeared, and he needed to figure out how to sell his candles online in a way that felt as personal as his in-store experience.

That's when David discovered an AI-powered marketing tool. It used data from his past sales to recommend what products people might want to buy next, helped him create personalized emails for his customers, and even suggested the best times to send promotions. The result? His online sales went through the roof, and he didn't need a team of marketers to do it.

David's story highlights something important: AI is not just for tech wizards or big corporations—it's for everyone. Whether you're looking to make your business more efficient, optimize your daily routine, or learn something new, AI is there to assist.

HOW CAN AI HELP ME TODAY?

Let's get specific. How can AI actually help you today? Here are a few real-world examples that show just how versatile and helpful AI can be:

1. Productivity Boost

Ever felt like there just aren't enough hours in the day to get everything done? AI tools can help you reclaim some of that time. Here's how:

Scheduling: AI-powered scheduling apps like Calendly or Google's smart assistant can automatically find the best times for meetings based on your availability, reducing the back-and-forth with colleagues or clients.

Emails: AI-driven email services can categorize your messages, suggest responses, and even summarize long email threads. Tools like Grammarly use AI to check your grammar and tone, making sure you always send polished, professional messages.

Project Management: Need help staying on top of deadlines? AI-driven project management tools like Trello or Asana can help you organize tasks, remind you of upcoming deadlines, and even allocate resources based on project needs.

2. Personal Assistant in Your Pocket

Imagine having a personal assistant available 24/7, but instead of paying a high salary, you simply open an app. AI-powered virtual assistants like Siri, Alexa, Google Assistant, or Cortana can make life easier by doing things like:

- Answering your questions ("What's the weather like today?")
- Setting reminders ("Remind me to call Mom at 4 p.m.")
- Automating tasks ("Turn off the lights at 10 p.m.")

It's like having a secretary who never sleeps, keeping you organized and on track. And the best part? You don't need to be tech-savvy to use these tools—they're designed to be user-friendly and intuitive.

3. Creative Collaboration

If you're an artist, designer, or musician, AI is opening up incredible possibilities. In fact, AI can be your creative partner:

Art & Design: Tools like DeepArt or Runway use AI to help artists create stunning visuals by learning from past art styles. You can input your own photos, and the AI will turn them into new, stylized artwork.

Music: AI-powered tools like Amper Music and AIVA can assist musicians in composing melodies or even suggest chord progressions and harmonies based on a few inputs. Imagine having an AI co-writer to jam with!

Writing: Writing can be tough, but AI tools like ChatGPT and Jasper can help you brainstorm, write drafts, and refine your text. Whether you're working on a novel or a business pitch, AI can help you get those ideas out faster.

AI IN HEALTHCARE: HELPING US LIVE BETTER, HEALTHIER LIVES

Nowhere is AI's potential more profound than in healthcare. In fact, the medical field is already using AI to improve outcomes for patients, make diagnoses faster, and develop new treatments.

Let me tell you about Linda, who was diagnosed with breast cancer. It was a terrifying time for her and her family. However, AI played a crucial role in her treatment. Her doctors used an AI-powered tool to analyze thousands of similar cases, finding patterns that led to a personalized treatment plan specifically designed for her. The AI considered everything—from her age to her genetics to the type of tumor she had—ensuring she got the most effective treatment possible.

In addition to helping doctors, AI is also helping patients directly. From fitness trackers that analyze your heart rate and sleep patterns to apps that provide mental health support, AI is playing a role in helping us lead healthier, happier lives.

WHY YOU CAN'T AFFORD TO IGNORE AI

The truth is, AI is becoming essential in almost every aspect of life. Whether it's enhancing your personal productivity, making your business more competitive, or even improving your health, AI is no longer a "nice-to-have" tool—it's a necessity.

And here's something to consider: the longer you wait to embrace AI, the harder it will be to catch up. *Think of AI like learning a new language*—the sooner you start, the more fluent you'll become over time. Those who start using AI today will have a significant advantage over those who wait until tomorrow.

BUT WON'T AI REPLACE JOBS?

It's impossible to talk about AI without addressing one of the most common fears: job loss. It's true—AI will disrupt the job market. But while some jobs will disappear, *new ones will be created*, and those who adapt will thrive.

Take the story of Melissa, a marketing consultant who worried that AI would make her job irrelevant. AI tools could analyze data and even write basic copy for ad campaigns. But instead of resisting, Melissa leaned in. She began using AI to enhance her work—automating time-consuming tasks like data analysis and focusing more on the creative aspects of her job. She didn't lose her job—she made it more efficient and, ultimately, more enjoyable.

This isn't an isolated story. Across industries, people are finding that while AI might change the way they work, it doesn't mean they'll be out of work. In fact, it might just make their jobs more rewarding by freeing them from repetitive tasks and allowing them to focus on what humans do best: creativity, empathy, and problem-solving.

AI AS YOUR NEW BEST FRIEND IN THE WORKPLACE

AI is the ultimate team player in the workplace. It can work alongside you, helping to identify new opportunities, streamline your workflow, and even offer insights that you may have overlooked.

For example:

Sales: AI can analyze customer data and behavior to help predict who's most likely to make a purchase, allowing sales teams to focus on high-potential leads.

Customer Service: AI-powered chatbots can handle routine customer inquiries, freeing up human agents to deal with more complex problems.

Human Resources: AI can assist in the recruitment process by scanning resumes and identifying the best candidates based on specific job criteria. This makes hiring more efficient while reducing human bias.

By working with AI rather than fearing it, you can become more effective, more productive, and more creative in your career.

CONCLUSION: AI AS YOUR ALLY

AI isn't something to be afraid of—it's something to embrace. Whether you're looking to enhance your personal life, become more productive at work, or simply keep up with the fast-paced world around you, AI is your ally.

In the next chapter, we'll explore *how to get started with AI*— whether you're a total beginner or someone looking to deepen your understanding. We'll cover everything from practical tools to resources you can use to start your AI journey with confidence.

Chapter 3

WHERE TO START

The AI Revolution
Taking the First Step

So, you're ready to embrace AI as your new friend, but you might be wondering: where do I begin? Starting something new can feel overwhelming, but with AI, the key is to take it one step at a time. You don't need to become a tech genius overnight. Think of it like learning a new hobby—you'll get better as you go, and soon enough, you'll wonder how you ever managed without it.

The most important thing to remember is that *AI is built to be user-friendly*. Whether you want to use it for work, creativity, learning, or personal productivity, there's an entry point for everyone.

IDENTIFY YOUR NEEDS

Before diving into the technical side of things, start by identifying *what you want AI to help you with*. This might seem simple, but it's a crucial first step. Think about your current life—what tasks take up the most time or energy? What areas of your personal or professional life could benefit from a little extra help?

Here are a few categories to consider:

Productivity: Are you looking for ways to streamline your work or manage your time better?

Creativity: Do you want a creative partner to help with writing, designing, or brainstorming ideas?

Learning: Are you interested in using AI to improve your skills or explore new topics?

Health and Well-being: Could you benefit from AI tools that track your fitness, mental health, or even suggest healthier habits?

Once you've identified your needs, you can start exploring specific tools that match your goals.

START SMALL: PRACTICAL TOOLS FOR BEGINNERS

If you're new to AI, the best advice is to *start small*. You don't have to dive into advanced AI programming or robotics to see the benefits. Here are some easy-to-use AI tools you can begin with, based on different areas of interest:

1. For Personal Productivity

Google Assistant / Siri / Alexa: Virtual assistants like these can help you manage everyday tasks—set reminders, schedule meetings, and even answer questions on the go.

Notion AI: Notion is a popular productivity tool that now integrates AI to help you organize your thoughts, create to-do lists, and even generate summaries of notes or meetings.

2. For Creativity

Canva: Canva is an easy-to-use design tool that offers AI features to help you create visually stunning graphics, presentations, and even social media posts.

ChatGPT: You can use AI to help with creative writing, from generating ideas to drafting blog posts, poems, or stories.

3. For Learning

Duolingo: Want to learn a new language? Duolingo uses AI to adapt to your learning pace and offers personalized exercises based on your progress.

Khan Academy: The Khan Academy platform is incorporating AI tutors to offer personalized learning experiences in subjects like math, science, and history.

4. For Health

MyFitnessPal: This app helps track your nutrition, workouts, and even sleep patterns. AI suggests personalized fitness routines and meal plans based on your goals.

Headspace: Headspace uses AI to offer personalized meditation practices that can help reduce stress, improve focus, and enhance your overall well-being.

5. Practice and Explore

Once you've started using AI tools, it's important to *experiment*. The more you use these tools, the more comfortable and confident you'll become. Take your time to explore different functions and see what works best for you. The beauty of AI is that it evolves with your needs.

For example:

- If you start using an AI-powered calendar app, notice how it learns your preferences over time. It might start suggesting the best times for meetings or remind you about tasks that fit your schedule.

- If you're working with creative AI tools like *MidJourney* (for generating artwork) or *Runway* (for editing videos), try different prompts or styles to see how the AI reacts.

You'll likely discover new ways to express your creativity that you hadn't thought of before.

Remember: *You don't need to be an expert on Day 1*. Every bit of practice helps you become more fluent with AI, just as you would with any other new skill.

SEEK OUT COMMUNITIES

One of the best ways to keep up with AI is to join communities of people who are also learning and experimenting. These communities can offer:

Tips and tricks for using different AI tools.

Guides and tutorials to help you get the most out of AI.

Support and encouragement when you hit a roadblock or feel unsure about something.

Communities can be found in many places:

Reddit: Subreddits like r/ArtificialIntelligence or r/MachineLearning offer great advice and resources for beginners.

Discord: Many AI-based apps or platforms have Discord communities where users can share their experiences, ask questions, and offer support.

YouTube: There are plenty of YouTube channels dedicated to AI tutorials, covering everything from basic introductions to advanced use cases.

By joining these communities, you'll not only stay updated but also make friends who are on the same AI journey as you.

HOW TO KEEP GROWING

The best part about AI is that it never stops improving, and neither should you! Here are a few ways to keep growing with AI once you've gotten the hang of things:

1. Stay Curious

Always be on the lookout for new AI tools that can make your life easier. The AI world is constantly evolving, and there are always new applications to explore.

2. Upgrade Your Skills

If you feel comfortable, consider learning more advanced AI skills like programming or data analysis. Platforms like *Coursera* or *Udemy* offer courses on AI that can help you dive deeper. Even a basic understanding of how AI works "under the hood" can give you more control over the tools you use.

3. Share Your Experience

The more you learn, the more you can help others. Don't be afraid to share your experiences with AI tools—whether it's in your workplace, with friends, or through online forums. You never know whose life you might improve by recommending the right AI solution!

FOR ADVANCED USERS:
TAKING YOUR SKILLS TO THE NEXT LEVEL

If you're already familiar with AI or have expertise in a specific field, the possibilities AI brings are even more exciting. For experienced professionals, creatives, or tech-savvy individuals, AI isn't just about automating tasks—it's a tool that can augment your capabilities and open up entirely new ways of working.

The key is to approach AI not as a replacement for your expertise but as an enhancement that allows you to achieve even greater things. In this section, we'll explore how AI can supercharge your existing skills in areas like data analysis, creativity, programming, and entrepreneurship.

ADVANCED AI TOOLS FOR ENHANCING EXPERTISE

For those with more experience, there's a whole world of AI tools designed to push the boundaries of what's possible. Here are some examples, tailored to different fields:

1. Data Science and Analysis

If you're working with large datasets or conducting research, AI can help you analyze information at a scale and speed that would be impossible manually.

TensorFlow and *PyTorch*: Both are open-source machine learning frameworks that allow you to build and train AI models. If you're into coding and data science, these are indispensable tools for creating deep learning models or neural networks to analyze complex data.

AutoML by Google: This tool helps automate the process of building machine learning models, allowing you to focus more on fine-tuning and less on repetitive tasks like feature engineering.

2. Creativity and Content Creation

For designers, musicians, writers, and other creative professionals, AI opens up entirely new realms of creativity:

Adobe Sensei: Integrated into Adobe Creative Cloud, Sensei uses AI to help with everything from photo editing to creating complex visual effects. For advanced users, you can use it to automate parts of your workflow, leaving you more time to focus on the finer details.

Runway ML: This tool allows artists and designers to experiment with machine learning in a creative environment. Whether you're generating music, creating generative art, or experimenting with video effects, Runway gives you the power to use AI in ways that push your creative boundaries.

3. Programming and Development

If you're a developer or tech enthusiast, AI is revolutionizing the way we write code and build software.

GitHub Copilot: This AI-powered coding assistant helps developers write code faster by suggesting whole lines or blocks of code based on the context. It can be especially useful when working on large-scale projects or when diving into unfamiliar programming languages.

OpenAI Codex: For developers working on AI-related projects, Codex (the AI behind GitHub Copilot) can help create more complex AI-powered applications. Codex can translate natural language prompts into code, making it a valuable tool for prototyping and speeding up development.

4. Entrepreneurship and Business Strategy

For entrepreneurs and business professionals, AI can be a game-changer when it comes to decision-making, marketing, and scaling your business.

ChatGPT Business Applications: AI chatbots and virtual assistants can handle customer support, gather data on consumer preferences, and even assist in marketing strategies by generating copy or analyzing market trends.

Tableau: For advanced business users, Tableau is a powerful data visualization tool enhanced by AI. It can help you turn raw data into actionable insights, perfect for improving decision-making in marketing, operations, and strategic planning.

AUGMENTING YOUR SKILLS:
COLLABORATION BETWEEN HUMAN AND AI

Once you've mastered the basics, it's time to think of AI as your **collaborator**. This is where the true power of AI lies for advanced users: it's not about replacing you but working alongside you to achieve more. Here are some ways to augment your expertise with AI:

1. Enhance Creative Output

For creative professionals, AI can serve as a brainstorming partner. Tools like *ChatGPT* can help generate ideas for novels, scripts, or visual content. It can give you a fresh perspective or even create variations on a theme you're working with.

Example: Let's say you're a screenwriter working on a sci-fi movie. You can use AI to generate plot twists, develop character backstories, or even help outline dialogue. While the AI won't write your masterpiece for you, it can certainly give you a fresh take when you're stuck.

2. Accelerate Data Processing

For data scientists or anyone dealing with large amounts of information, AI can help process and interpret data faster. Rather than spending hours or days combing through datasets, you can use AI to pre-process the data and surface insights that may otherwise go unnoticed.

Example: Imagine you're working asa a financial analyst for a global company. AI can scan through hundreds of financial reports and suggest key trends, allowing you to make more informed decisions quickly.

3. Amplify Marketing Strategies

For marketing professionals, AI tools can take your strategy to the next level. AI can help you analyze customer data, predict trends, and even automate content creation at scale.

Example: If you run a digital marketing agency, AI can help you create hyper-personalized ad campaigns by analyzing customer behavior. With platforms like *HubSpot's AI-powered marketing suite*, you can automate email campaigns and optimize ads based on real-time data analysis.

4. Unlock New Levels of Problem Solving

In industries like engineering or scientific research, AI can help you solve problems that are too complex for traditional methods.

Example: If you're working in drug development, AI can speed up the process of discovering new medicines by analyzing millions of chemical compounds and predicting which combinations might work best.

STAYING AHEAD OF THE CURVE: ADVANCED LEARNING OPPORTUNITIES

If you're already familiar with AI but want to stay ahead of the curve, there are plenty of ways to continue your learning journey. Here are a few tips for those who are looking to master AI on a deeper level:

1. Master New Frameworks and Languages

AI evolves quickly, and so do the tools and languages used to build it. If you want to stay ahead in the tech world, make sure you're keeping up with the latest frameworks. Platforms like *Coursera*, *edX*, and *Udacity* offer specialized courses in AI, data science, and machine learning.

2. Specialize in a Niche Area

AI has applications in almost every field, but if you want to be an expert, consider specializing in a niche area like natural language processing (NLP), computer vision, or AI ethics Focusing on a specific aspect of AI can make you an invaluable asset in your industry.

3. Contribute to AI Development

If you're already an advanced AI user, consider contributing to the community. Many AI tools and platforms are open source, meaning anyone can contribute to their development. By helping to improve AI systems, you not only grow your skills but also give back to the AI community.

CONCLUSION: YOUR AI JOURNEY STARTS NOW

Starting with AI doesn't have to be complicated or intimidating. It's about taking small steps, using the right tools, and staying curious. AI is here to assist you, and the more you engage with it, the more it will enhance your life.

In the next chapter, we'll explore how to **prepare for AI's ever-evolving future**, ensuring that you stay relevant and empowered in this rapidly changing landscape.

Chapter 4

PREPARATION

The AI Revolution
Building a Futureproofing Mindset

As AI continues to evolve, preparing for the future requires more than just learning new tools. It's about cultivating the right mindset, building adaptable skills, and aligning ourselves with the changing landscape. In this chapter, we'll dive into the ways we can prepare to not only survive but thrive in the AI revolution.

EMBRACING A GROWTH MINDSET

Preparation starts with how you perceive change. AI technology is moving fast, but your ability to keep up is directly tied to your mindset. What separates those who thrive from those who struggle is often not just skill but attitude. A *growth mindset*, a concept developed by psychologist Carol Dweck, is the belief that abilities and intelligence can be developed with time and effort. When applied to AI, this means believing that you are capable of understanding and working with AI, no matter where you start.

How to Foster a Growth Mindset:

View challenges as opportunities: When you encounter new AI tools or concepts that feel daunting, see them as chances to grow. Each challenge is an opportunity to expand your knowledge.

Embrace mistakes as learning experiences: Mistakes are not a sign of failure but of progress. The more you experiment, the more you learn.

Celebrate small wins: Each step forward, no matter how small, is worth celebrating. Whether you've just learned how to use a simple AI tool or completed a more complex project, take a moment to acknowledge your growth.

By cultivating a growth mindset, you set the foundation for adapting to the rapid advancements AI brings. This will give you the confidence to tackle whatever comes next with curiosity rather than fear.

INVESTING IN LIFELONG LEARNING

To prepare for the AI-driven future, continuous learning is crucial. AI is a field that evolves constantly, with new breakthroughs happening all the time. But the great news is that there are countless resources to help you stay up to speed, from online courses to community groups and open-source platforms. Lifelong learning is not just about mastering AI—it's about being open to new ideas, approaches, and perspectives.

How to Invest in Lifelong Learning:

Set aside time regularly: Block out time each week to learn something new, whether it's taking an AI-related course, reading an article, or experimenting with a new tool.

Focus on your interests: You don't need to learn everything at once. Focus on the areas of AI that most excite you—whether it's how AI can improve healthcare, revolutionize art, or transform business.

Engage with communities: Join forums, attend webinars, or participate in AI-related groups. Being part of a learning community can boost your motivation and help you stay informed.

Lifelong learning is not just a way to keep up with AI—it's a way to stay energized, curious, and ahead of the curve.

RESILIENCE IN THE FACE OF CHANGE

Change, especially at the pace AI is developing, can be overwhelming. But resilience is the ability to bounce back from challenges and adapt to new realities. Resilience means being ready to navigate the uncertainty of AI's future, whether that involves learning new skills, adapting to different ways of working, or even facing disruptions in traditional industries.

How to Build Resilience:

1. *Embrace uncertainty*: Accept that change is a constant, especially in the world of AI. By seeing uncertainty as a natural part of growth, you become more flexible and less fearful.

2. *Practice self-compassion*: Be kind to yourself as you navigate the AI landscape. It's okay to feel unsure or frustrated at times. Self-compassion helps you stay motivated rather than discouraged.

3. *Create a support network*: Surround yourself with a community of learners, mentors, or peers who are also navigating AI. Sharing experiences can make you feel more connected and resilient.

Resilience ensures that when you encounter roadblocks, you'll have the strength and mindset to keep going, making sure you don't fall behind but continue progressing toward your goals.

DEVELOPING ADAPTABLE SKILLS

In the AI era, some skills are more important than ever, especially those that make you adaptable. The future isn't just about technical skills; it's about soft skills like creativity, critical thinking, and emotional intelligence. These qualities are uniquely human and are areas where AI cannot fully replace us.

Key Adaptable Skills to Cultivate:

1. *Creativity*: AI can assist in generating ideas, but human creativity is irreplaceable. Cultivate your creative thinking by experimenting, thinking outside the box, and merging disciplines (e.g., combining art and technology).

2. *Critical Thinking*: With so much information available, the ability to critically assess AI-generated insights and make informed decisions is key. Question the data, analyze it, and use your judgment to guide AI-driven decisions.

3. *Emotional Intelligence*: AI can analyze emotions, but it can't replace human empathy. In leadership roles, managing teams, or interacting with clients, emotional intelligence will set you apart. It will also help you better collaborate with AI, by understanding how and when to apply its capabilities.

The combination of technical knowledge and these adaptable skills will future proof your career and personal development, allowing you to evolve alongside AI rather than be left behind.

CREATING A PERSONALIZED AI LEARNING PLAN

Now that we've discussed the importance of mindset, learning, resilience, and adaptability, it's time to create a personalized AI learning plan. This plan should be tailored to your current knowledge, interests, and long-term goals. Start small but stay consistent.

Steps to Create Your AI Learning Plan:

1. *Assess your current skills*: Start by identifying what you already know about AI. Are you a beginner, intermediate, or advanced user? What are your strengths and weaknesses?

2. *Identify your goals*: What do you want to achieve with AI? Are you looking to use AI in your business, improve your personal productivity, or maybe even develop AI-related projects?

3. *Choose your learning path*: Based on your goals, decide what to focus on. For beginners, this might involve understanding AI basics like machine learning or natural language processing. For more advanced learners, you may want to explore specific frameworks or industry applications.

4. *Schedule regular learning sessions*: Consistency is key. Set 3. *Choose your learning path*: Based on your goals, decide what to focus on. For beginners, this might involve understanding AI basics like machine learning or natural language processing. For more advanced learners, you may want to explore specific frameworks or industry applications.

aside specific times each week to learn and practice AI skills. Use online courses, books, or even tutorials on platforms like YouTube.

5. *Stay flexible and open to change*: As you progress, your goals may shift, and that's okay. Stay flexible and adjust your learning plan as you discover new opportunities or interests.

PREPARING FOR AI IN THE WORKPLACE

For many, AI is already reshaping the workplace. Whether you're a business owner, an employee, or a freelancer, it's crucial to prepare for how AI will impact your career or industry. In the coming years, jobs will evolve, some roles will disappear, and new ones will emerge. The key is being ready for this shift and embracing AI as a co-worker.

How to Prepare for AI in Your Workplace:

1. *Understand the AI tools in your industry*: Every industry is being touched by AI in different ways. If you're in healthcare, it might be AI for diagnostics. In marketing, AI for data analysis. Get familiar with the specific AI tools that are reshaping your field.

2. *Work alongside AI, not against it*: Instead of fearing that AI will take your job, think about how it can help you perform better in your role. Can AI handle routine tasks, giving you more time for strategic work? How can you leverage AI to achieve better results?

3. *Prepare for new job roles*: As AI changes the workforce, new job titles will emerge. Keep an eye on trends and be ready to upskill or shift roles as needed.

Being proactive about how AI will affect your work will give you a competitive edge, ensuring that you are not left behind as industries change.

FUTUREPROOFING YOUR CAREER AND LIFE

As you prepare for the AI revolution, remember that this is about more than just career advancement. It's about future-proofing your entire life. The skills, mindset, and adaptability you develop will serve you not only in your job but in personal growth and life satisfaction.

Whether you're looking to thrive in the workplace, pursue a passion project, or simply navigate the changing world with confidence, the preparation you put in now will make all the difference.

CONCLUSION: PREPARATION AS A LIFELONG JOURNEY

Preparing for AI is not a one-time task—it's a lifelong journey. By embracing a growth mindset, investing in continuous learning, building resilience, and developing adaptable skills, you'll be well-equipped to thrive in the AI era. More than just keeping up with technology, this preparation is about ensuring that you can shape your future on your terms.

In the next chapter, we'll explore how AI is transforming the world of art and creativity, and how you can use AI to enhance your own creative expression.

Chapter 5

ARTISTS AND AI

The AI Revolution
Collaborating with Creativity

The world of art, once seen as a purely human domain, is now being transformed by AI. Whether through music, visual arts, or literature, AI is becoming a powerful tool for artists, not as a replacement but as a collaborator. This chapter delves into the exciting intersection of AI and creativity, exploring how artists can embrace AI to expand their creative horizons.

THE EVOLUTION OF CREATIVE TOOLS

Throughout history, advancements in technology have influenced the way artists create. From the invention of the camera to digital painting software, artists have always found ways to merge their creativity with new tools. AI is the next step in this evolution—a tool that doesn't just execute but also learns and creates alongside the artist.

What makes AI different from other tools is its ability to *generate*, *suggest*, and *enhance*. Rather than merely providing a platform for the artist, AI can now actively participate in the creative process. But far from threatening the role of human creativity, this partnership opens up new possibilities, allowing artists to push the boundaries of their imagination.

AI AS A CREATIVE ASSISTANT

Think of AI as a collaborator that helps you with creative brainstorming, providing new ideas and perspectives you may not have thought of on your own. Many AI tools can analyze patterns, colors, compositions, and even artistic styles, giving artists suggestions on how to evolve their work.

WAYS AI CAN ASSIST IN ART

Idea Generation: AI can help artists generate new concepts or themes based on existing work. Tools like *DALL·E* or *DeepDream** generate original images from text or inputted data, allowing visual artists to explore new directions.

Style Transfer: AI can take one artistic style and apply it to another. Imagine being able to paint in the style of Van Gogh or Picasso at the click of a button, then combining it with your own unique touch.

Music Composition: For musicians, AI tools like *Amper* or *AIVA* can compose music based on predefined inputs, whether it's creating beats, melodies, or full orchestral compositions. This helps musicians experiment with genres or arrangements that they might not typically explore.

Writing and Poetry: AI tools like *ChatGPT* can assist writers and poets in generating text, offering suggestions for dialogue, plot twists, or even poetic structures.

Rather than replacing the artist, AI acts as a *muse*, inspiring new ideas, offering options, and pushing boundaries in ways that might not be possible alone.

BREAKING THROUGH CREATIVE BLOCKS

Every artist faces creative blocks. Whether it's a blank canvas, a stuck melody, or an unfinished manuscript, these moments can be frustrating.

AI can help break through these blocks by suggesting new ideas, offering fresh directions, or simply providing an alternative perspective.

How AI Helps Overcome Creative Blocks:

1. *Generating Random Ideas*: AI can help you move past stagnation by generating ideas based on your input. By presenting unexpected results, it forces you to look at your work from a new angle.

2. *Providing Variations*: AI can produce multiple variations of an idea in seconds. For example, if you're a visual artist unsure about the color scheme, AI can provide different palettes and layouts, giving you more options to choose from.

3. *Offering Inspiration*: Sometimes, you just need a spark of inspiration. AI can generate images, texts, or melodies based on different parameters, serving as a jumping-off point for your creativity.

By taking over some of the heavy lifting, AI allows artists to focus on the essence of their creativity, enabling them to work with greater flow and less friction.

ELEVATING SKILLS WITH AI

While AI can assist in generating content, its greatest potential lies in helping artists elevate their existing skills. Rather than being a shortcut, AI tools can *enhance* an artist's abilities by automating tedious tasks, allowing more time for creative exploration and refinement.

Examples of How AI Can Enhance Artistic Skills:

Refining Techniques: AI-powered software can analyze your work, identify areas for improvement, and offer suggestions.

This is especially helpful in fields like photography, where AI tools can adjust lighting, focus, and composition to help create more polished results.

Enhancing Precision: In disciplines like digital painting or 3D modeling, AI tools can assist in creating detailed, precise elements that would take hours to manually render, such as complex textures or lighting effects.

Speeding Up the Creative Process: AI can take on repetitive or time-consuming tasks, like rendering or editing, allowing you to focus more on the creative aspects of your project.
For those artists who feel intimidated by learning advanced software or technology, AI acts as a friendly guide, helping to streamline processes while giving artists more freedom to explore their creativity.

COLLABORATING WITH AI – A NEW CREATIVE PARTNER

Many artists today are embracing the idea of *co-creation* with AI. AI is not simply a tool you use, but a creative partner you can collaborate with. Whether you are a visual artist, writer, or musician, AI can take on certain aspects of the creative process, while you guide the vision.

For example, some painters work with AI to generate base compositions and then add their personal touch. Musicians can let AI generate background scores and then compose the lead melodies themselves. Writers might use AI to generate the structure of a story and then infuse it with their own style and voice.

In this sense, AI doesn't replace the artist but becomes an extension of the artist's creative potential.

THE ETHICAL DILEMMAS OF AI IN ART

As with any new technology, the rise of AI in creative fields brings with it a set of ethical considerations.

For instance, some question whether AI-generated works can truly be considered "art" in the traditional sense.

Others worry about how AI-generated art might impact the livelihoods of artists, particularly when AI-generated pieces start competing in the same commercial spaces as human-created works.

Key Ethical Questions to Consider:

- *Who Owns AI-Generated Art?*: If an artist uses AI to create a piece, who holds the copyright? The artist, the AI developer, or both?

- *Is AI-Generated Art Original?*: AI relies on patterns and data from existing works. Does this mean AI-generated art is derivative, or can it be considered original in its own right?

- *Impact on Jobs*: Will AI-generated art replace human artists, or will it simply create new opportunities for collaboration and innovation?

While these questions don't have clear answers yet, it's important for artists to be aware of the ethical landscape and think about how they want to engage with AI in a way that honors their craft and values.

PUSHING THE BOUNDARIES OF HUMAN IMAGINATION

One of the most exciting aspects of AI in the arts is its ability to push the boundaries of what we think is possible. AI can come up with combinations, patterns, and structures that may be beyond human imagination—offering artists entirely new directions.

Areas Where AI Pushes Creative Boundaries:

Generative Art: Artists use AI to create generative art, where algorithms generate art based on a set of parameters.

This allows for the creation of infinitely different pieces of work, all based on a core idea.

AI in Interactive Art: AI can create art that interacts with the audience in real-time. For example, installations where the audience's movements or sounds directly influence the art being created in the moment.

AI and Virtual Reality (VR): Combining AI with VR opens up immersive worlds where users can interact with evolving art in ways never before imagined.

AI offers the opportunity to expand our creative horizons, blending human intuition and creativity with machine precision and computational power.

BECOMING A HYBRID ARTIST

In the AI era, artists are becoming more than just traditional creators—they are evolving into *hybrid artists*, blending technology with creativity. This new wave of creators is comfortable using AI to enhance their artistic process, while still holding onto the core of human creativity.

To become a hybrid artist, you don't need to be a programmer or a data scientist. It's about learning how to *collaborate* with AI, making it part of your creative toolkit. Artists who embrace this hybrid approach will be at the forefront of the next artistic revolution.

AI IN VISUAL ARTS: FRIEND OR FOE?

For visual artists, AI presents both exciting opportunities and real concerns. While AI can serve as a creative partner, there's an underlying fear that it might undermine the value of human-made art. Many artists worry that AI-generated content could flood the market, devaluing original works or making it harder for human artists to stand out.

Key Concerns for Visual Artists:

1. *Loss of Authenticity*: One major concern is that AI-generated art lacks the human touch—the emotion, intention, and experience that go into creating an original piece. Artists often ask, "If anyone can use AI to create art, where is the value in human creativity?"

2. *Copyright Issues*: Since AI often learns from existing artworks, there are concerns about whether it could unintentionally copy or infringe on the work of human artists. Should artists be credited when AI uses their style or themes to create new works?

3. *Market Saturation*: As AI becomes more common in art creation, there is a risk of oversaturation in the market. Will collectors and audiences still seek out unique, human-made art when AI can produce high-quality pieces faster and cheaper?

NAVIGATING THE ETHICAL LANDSCAPE

Artists are also grappling with ethical questions about AI's role in the creative process. Many are wondering if AI-generated works are *original* or if they're simply remixing the past. Additionally, what role should the artist play in guiding the output of AI? If an artist collaborates with AI, does it diminish their authorship of the work?

Ethical Considerations for Artists:

Attribution: Should AI be credited as a co-creator? Should artists disclose the use of AI in their works, and how does this affect the perception of authenticity?

Cultural Sensitivity: There are concerns about AI unintentionally perpetuating cultural biases or appropriating certain artistic styles without understanding their cultural significance. Artists working with AI need to be aware of how their inputs may impact the broader cultural conversation.

Ownership: If an AI is trained on public artworks, who owns the output? This is a contentious issue, particularly in fields like photography and design, where copyright laws can become blurry.

A Way Forward:

Transparency and Accountability: One-way artists can address these concerns is by being transparent about their collaboration with AI. If AI plays a significant role in a piece, acknowledging it might not detract from the art but rather open up new conversations about creativity in the digital age.

Embracing Uniqueness: While AI may become more prominent in creating art, the unique *vision* and *perspective* that each artist brings to the table can never be replaced. AI cannot replicate the personal stories, emotions, and life experiences that drive human creativity. Artists should focus on using AI as a tool to *enhance* their voice, not replace it.

MUSICIANS AND AI: REVOLUTION OR REDUNDANCY?

For musicians, AI is also a double-edged sword. On one hand, it can automate certain aspects of composition and production, freeing up time for more creative pursuits. On the other, there's a fear that AI could replace human musicians or lead to a homogenization of music.

AI IN MUSIC CREATION

AI has made significant strides in music production, with tools like *AIVA* and *Amper* composing melodies, harmonies, and even entire tracks based on input data. Musicians can collaborate with AI to experiment with new genres or create music faster than ever before.

Opportunities for Musicians:

1. *Enhanced Composition*: AI can help musicians compose music by analyzing vast amounts of data, from classical compositions to contemporary hits. By identifying patterns in rhythm, harmony, and melody, AI can suggest new chord progressions or help craft unique arrangements.

2. *Remixing and Sampling*: AI is also useful for remixing and sampling existing tracks. Musicians can use AI to quickly rework existing compositions, giving old tracks new life while maintaining their creative control.

3. *Production Speed*: AI can streamline the music production process by automating time-consuming tasks, such as mixing and mastering, allowing musicians to focus on the creative aspects of their work.

CHALLENGES FOR MUSICIANS

1. *Loss of Human Emotion*: While AI can generate music, many musicians argue that it lacks the *emotional depth* and nuance that come from a human performer. Music is often about connection and storytelling—things that AI may struggle to truly capture.

2. *Job Displacement*: There's also concern that AI could replace session musicians, composers, and producers, particularly in industries like advertising or film scoring, where quick turnaround and cost efficiency are priorities. How will human musicians maintain relevance in an industry increasingly relying on algorithms?

3. *Homogenization of Sound*: One potential downside of relying heavily on AI is the risk of *homogenization*—where music starts to sound more similar because it is generated based on existing patterns and trends. This could limit innovation and diversity in music if not carefully managed.

ETHICS IN MUSIC CREATION

Attribution and Rights: Just as in the visual arts, musicians need to consider how much credit should be given to AI when it plays a significant role in composing or producing a track. Should the AI or its developers be credited? How will royalties be distributed for AI-assisted compositions?

Cultural Sensitivity: Musicians must also be mindful of how AI samples and integrates music from different cultures. There's a risk that AI could appropriate styles or sounds without understanding their cultural significance.

COLLABORATION VS. COMPETITION
A FUTURE FOR ARTISTS AND MUSICIANS

Rather than viewing AI as competition, many artists and musicians are finding ways to *collaborate* with it. Some musicians are using AI to enhance their songwriting or production, while others are creating entirely new genres that blend human creativity with machine learning.

Ways to Collaborate with AI in Music:

Interactive Compositions: Some musicians are experimenting with real-time collaborations with AI, where the AI responds to live performances by generating complementary melodies or rhythms. This creates a dynamic, interactive performance experience.

AI-Driven Live Performances: Musicians like *Taryn Southern* have used AI to co-produce entire albums. Southern's album "I AM AI" is an example of how artists can leverage AI tools to compose music that blends human emotion with machine precision.

Creating New Sounds: Musicians are also using AI to generate entirely new sounds, creating instruments or effects that have never been heard before. By training AI on a wide array of sound data, musicians can produce innovative sonic landscapes.

For visual artists, the opportunity to collaborate with AI lies in pushing the boundaries of traditional mediums. AI can assist in creating larger-scale works, conceptualizing new forms, or even helping artists work across mediums, blending music, visual arts, and performance in ways previously unimaginable.

PRESERVING THE HUMAN TOUCH

Despite the growing presence of AI in the creative world, there will always be a place for *the human touch*. Art and music are not just about technical mastery but about conveying emotions, experiences, and stories. AI can assist with the technical side of creation, but it is the *human spirit* that gives art its meaning.

Strategies for Maintaining the Human Element in Art and Music

1. *Keep Emotion at the Core*: No matter how advanced AI becomes, it cannot replicate the personal, emotional experiences that inspire art. Artists and musicians should continue to draw from their personal lives, cultures, and emotions to ensure their work remains authentic.

2. *Use AI as a Springboard*: Rather than letting AI dominate the creative process, artists can use it as a *springboard*—a way to get new ideas flowing or to enhance existing work, but not to take over the entire creative journey.

3. *Focus on Innovation*: Artists and musicians who embrace experimentation and innovation will thrive in the AI era. Whether it's combining new sounds, exploring different art styles, or blending technology with tradition, those who push boundaries will continue to stand out.

A NEW RENAISSANCE FOR ARTISTS AND MUSICIANS?

Many believe that AI could spark a *new renaissance* in the arts. By freeing artists and musicians from some of the more technical aspects of their work, AI allows them to focus more deeply on their vision, creativity, and innovation. Far from replacing human creativity, AI could usher in an era of *unprecedented artistic freedom*.

VISION FOR THE FUTURE

Interdisciplinary Art: Artists and musicians may find themselves blending more disciplines as AI enables new forms of collaboration between visual art, music, dance, and performance.

Customization at Scale: AI allows artists to tailor their work for different audiences. Imagine a musician composing a song that adapts to the listener's mood, or an artist creating a piece that changes based on the viewer's location or time of day.

Global Artistic Communities: AI can help artists and musicians from around the world collaborate, creating truly global art forms. By breaking down language and geographical barriers, AI fosters a new kind of artistic *collective creativity*.

CONCLUSION: AI AS A MUSE, NOT A REPLACEMENT

As we've explored in this chapter, AI is not a threat to creativity—it's a new muse. Artists who learn to collaborate with AI will find that their creative potential is expanded, not diminished. The future of art lies in the partnership between human creativity and machine intelligence, pushing boundaries in ways we've never seen before.

In the next chapter, we'll dive into how AI is reshaping the job market and what you can do to stay competitive in an increasingly AI-driven world.

Chapter 6

AI AND JOBS

The AI Revolution
Navigating the Future of Work

The rise of AI has sparked widespread conversations about the future of work, leading to both excitement and anxiety. While AI has the potential to create new job opportunities and revolutionize industries, there's also the fear of job displacement, particularly in fields that rely on repetitive or data-driven tasks.

THE CURRENT LANDSCAPE OF AI IN THE WORKFORCE

In today's world, AI is already making significant inroads into various industries. From automating customer service with chatbots to optimizing supply chains, the use of AI is transforming business operations and how we approach work. Jobs in industries like healthcare, finance, manufacturing, and retail are undergoing massive shifts as AI tools improve efficiency and accuracy.

THE FEAR OF JOB LOSSES

One of the most common concerns is that AI will replace human workers, especially in industries where routine tasks can easily be automated. This fear is not unfounded—many jobs, particularly in manufacturing, transportation, and administrative roles, are already being affected by AI-driven automation. For example:

Manufacturing: AI-powered robots are taking over tasks that used to be performed by human assembly line workers.

Transportation: Self-driving vehicles, while still in development, have the potential to disrupt jobs in trucking, delivery services, and even ride-hailing.

Administrative Jobs: AI systems can automate data entry, scheduling, and other clerical tasks, leading to fewer jobs in office administration.

While these changes are causing understandable anxiety, it's important to acknowledge that AI is also creating new types of jobs—roles that didn't exist just a few years ago.

THE EMERGENCE OF NEW JOB OPPORTUNITIES

AI isn't just eliminating jobs; it's also *creating new opportunities* across various sectors. As the technology advances, a need for workers with specialized skills to manage, interpret, and implement AI systems grows. These new roles often require a mix of technical knowledge and creative thinking, offering new paths for professional growth.

JOB OPPORTUNITIES IN THE AI ERA

AI Specialists: Professionals who understand AI programming, machine learning, and data science are in high demand. These roles involve designing and developing AI systems or improving existing ones.

Data Analysts: With AI generating massive amounts of data, companies need skilled analysts to interpret and make sense of this information, turning it into actionable insights.

AI Trainers: AI systems often require human trainers to teach them how to interpret data accurately. For instance, in natural language processing, AI trainers are needed to help systems like chatbots understand the nuances of human language.

Ethical AI Consultants: As AI becomes more prevalent, the need for ethical oversight increases. Companies are hiring professionals to ensure that AI systems are being used in ways that are fair, transparent, and non-discriminatory.

AI-Augmented Roles: In many professions, AI serves as a tool to enhance human capabilities rather than replace them. For example, doctors use AI to assist with diagnoses, and marketers use AI to analyze consumer behavior and predict trends.

While we've touched on industries like healthcare, finance, and retail, many other sectors will also be reshaped by AI. Here are some additional fields where AI will likely have a profound impact:

Legal Services: AI is already being used to analyze contracts, sift through vast amounts of legal data, and even predict the outcomes of court cases. While this could make legal services more accessible, jobs for paralegals, legal researchers, and even junior attorneys may be at risk.

   Opportunities: AI could streamline routine legal tasks, allowing lawyers to focus on strategy and client relations. It may also enable firms to take on more cases by automating parts of the process.

   Challenges: Workers in entry-level legal roles may need to reskill, transitioning to roles like AI system management or client-facing responsibilities.

Agriculture: AI-driven machines are revolutionizing farming by automating everything from planting and watering crops to monitoring soil health. While this can boost efficiency, it might also reduce the need for traditional farming labor.

   Opportunities: AI can optimize yield, reduce waste, and make farming more sustainable. Roles like agricultural data analysis and AI-driven machinery maintenance could emerge.

Challenges: Traditional farmers may need to learn how to work with AI-powered systems or shift their focus to overseeing and managing these technologies.

Marketing and Advertising: AI is transforming how companies engage with consumers, from personalized ads based on browsing habits to AI-driven content creation. While this enhances targeted marketing, it might reduce the need for traditional marketing roles.

Opportunities: AI allows marketers to better understand consumer behavior, creating highly personalized campaigns that resonate more deeply. New roles in AI-driven strategy and content optimization will emerge.

Challenges: Roles focused on manual data analysis and repetitive tasks in marketing might be automated, requiring professionals to adapt to AI-enhanced strategies.

Construction: In construction, AI can assist with project management, site safety, and predictive maintenance. Drones powered by AI are used for surveying sites, and robots are starting to take on tasks like bricklaying or pouring concrete.

Opportunities: AI will improve efficiency, safety, and cost management. Jobs in AI-driven project management and construction robotics are on the rise.

Challenges: Manual labor jobs in construction may be replaced or enhanced by automation, leading to shifts in how workers contribute to projects.

PREPARING FOR THE SHIFT: ADAPTABILITY IS KEY

One of the most important qualities workers can cultivate in the age of AI is *adaptability*. As industries evolve, so too must the skill sets of the workforce. Continuous learning, flexibility, and a willingness to embrace new technologies are crucial for thriving in a world where AI plays a central role.

DEVELOPING NEW SKILLS

To stay competitive in the AI-driven job market, workers must be proactive about learning new skills that complement the technology. This doesn't necessarily mean becoming an AI expert, but rather learning how to work alongside AI tools effectively.

Key Skills to Develop:

1. *Digital Literacy*: Having a basic understanding of how AI and other technologies work is essential in almost every field. Workers should familiarize themselves with the tools and platforms used in their industry to stay relevant.

2. *Creative Problem-Solving*: AI excels at handling routine tasks, but it often struggles with creativity and critical thinking. Workers who can solve complex problems, think outside the box, and come up with innovative solutions will continue to be in high demand.

3. *Emotional Intelligence*: While AI can automate many tasks, it cannot replace the human capacity for empathy, communication, and leadership. Emotional intelligence will become even more valuable in fields like customer service, healthcare, and education, where human connection is key.

4. *Collaboration and Teamwork*: AI can assist with tasks, but it's still up to humans to collaborate, make decisions, and drive projects forward. Those who can work well in teams and leverage AI to improve efficiency will thrive in the workplace of the future.

LIFELONG LEARNING AND UPSKILLING

As AI continues to evolve, so too will the skills required in the workforce. *Lifelong learning* is no longer optional; it's a necessity. Workers need to regularly upskill to keep pace with the changes brought on by AI.

Many companies are already recognizing this and offering *reskilling* programs for their employees. These programs are designed to teach workers new skills that are relevant to AI-driven industries, ensuring they remain employable as technology advances.

INDUSTRIES MOST AFFECTED BY AI

Certain industries are likely to feel the impact of AI more than others. While some sectors may see job displacement due to automation, others will see an increase in demand for skilled workers.

Healthcare:

AI is transforming healthcare by assisting with diagnosis, treatment plans, and even surgery. While AI can automate tasks like reading medical images, it also enhances the ability of doctors and nurses to provide better care.

Opportunities: AI-assisted medical devices, personalized treatment plans, AI diagnostics.

Challenges: Job displacement for administrative roles, need for reskilling of healthcare workers.

Finance:

In finance, AI is streamlining operations, enhancing fraud detection, and improving customer service. However, jobs in financial analysis, trading, and banking are at risk of being automated.

Opportunities: AI-driven financial services, robo-advisors, fraud prevention.

Challenges: Job loss in traditional financial roles, need for upskilling in AI-driven tools.

Retail:

AI is reshaping the retail industry by improving inventory management, personalizing customer experiences, and enabling more efficient supply chains. However, traditional retail jobs, such as cashiers and stock clerks, are at risk.
Opportunities: AI-enhanced customer service, personalized marketing, automated inventory.

Challenges: Displacement of low-skill jobs, need for employees with AI management skills.

REFRAMING THE FUTURE: A PARTNERSHIP WITH AI

Rather than fearing AI, workers can benefit by embracing it as a *partner* in their professional lives. AI is not here to replace human ingenuity, but rather to *enhance* it. The key lies in reframing AI as a tool that can free workers from mundane, repetitive tasks, allowing them to focus on higher-level thinking, creativity, and strategy.

A BALANCED FUTURE: HUMANS AND AI TOGETHER

The future of work will require balance. As AI takes over more routine and data-driven tasks, humans will have the opportunity to focus on areas that require a *human touch*—creativity, empathy, leadership, and innovation. Rather than fearing AI, workers should view it as an enabler of human potential, providing the tools to take work to new heights.

As industries transform and the workforce adapts, it's clear that AI will play a central role in the future of work. However, the future is not one of displacement, but of *collaboration*. Workers who develop the skills to work alongside AI will thrive in this new era.

THE ROLE OF EDUCATION AND TRAINING INSTITUTIONS

Educational institutions must also evolve to prepare students for the AI-powered future. Traditional models of education, which emphasize memorization and standardized testing, may not be enough to equip students with the skills they'll need in an AI-driven workforce.

The way children are educated will be dramatically reshaped by AI, with both exciting opportunities and challenges.

AI-ENHANCED LEARNING

AI has the potential to revolutionize education by personalizing the learning experience for every child. Currently, students are taught in large groups, with a "one-size-fits-all" approach. But AI can tailor education to fit each child's unique needs, strengths, and interests.

Personalized Learning Paths:

AI can create individualized learning plans for each student, identifying their strengths and weaknesses and adjusting lessons accordingly. Imagine your boys each receiving lessons tailored specifically to their pace, interests, and needs. If one excels in math and the other in creativity, AI can adapt to nurture those areas. Interactive Learning:

AI-powered platforms can offer interactive, engaging lessons that go beyond traditional textbook learning. Gamified learning, virtual reality classrooms, and AI tutors could make education far more engaging and effective.

Access to Global Knowledge:

AI can bridge the gap between students from different parts of the world by offering access to the same resources, making learning more equitable. Young learners could learn from the best educators around the globe, accessing a wealth of knowledge from different cultures, languages, and perspectives.

CHALLENGES AND CONSIDERATIONS

While the benefits of AI in education are immense, there are also concerns that parents may have:

Over-reliance on Technology: There is a risk that children could become overly dependent on AI tools, potentially hindering the development of critical thinking and problem-solving skills. As a parent, striking a balance between AI-assisted learning and traditional forms of education will be crucial.

Privacy and Data Security: AI-driven educational tools often require vast amounts of data to function effectively. Ensuring the privacy and security of children's data will be a top priority for schools and parents alike. Young learners' information, like their learning habits and progress, must be protected.

Human Connection: While AI can deliver personalized education, it cannot replace the importance of human teachers and mentors who provide emotional support, encouragement, and social interaction. AI will enhance the learning experience, but real-life teachers will remain essential.

PREPARING CHILDREN FOR AN AI-DRIVEN WORLD

As a parent, one of your primary concerns is ensuring that your children are prepared for the world they'll inherit. Here are some ways you can help them get ready for the AI-driven future:

Encourage Curiosity and Lifelong Learning: AI will evolve rapidly, and the skills they learn today may be outdated by the time they're adults. Instilling a love of learning and a curiosity about the world will be invaluable. Teach them to embrace change and seek out new knowledge.

Promote Problem-Solving Skills: While AI can assist with many tasks, creative thinking and problem-solving will always be crucial human skills. Encourage young learners to tackle challenges, think outside the box, and come up with innovative solutions—skills that will serve them in any field.

Balance Technology with Hands-On Experiences: While AI tools will be a big part of their lives, make sure young learners also have plenty of opportunities for hands-on, real-world experiences. Encourage them to build, create, and explore the world beyond screens, ensuring they develop a broad range of skills.

Teach Empathy and Emotional Intelligence: AI may be able to mimic human interactions, but it cannot replace genuine human emotions and connections. Teaching young learners to be kind, empathetic, and emotionally aware will help them thrive in an AI-driven world where human-centered skills are highly valued.

THE ROLE OF AI IN SHAPING THE CLASSROOM

Let's also look at how the traditional classroom itself will change with AI integration.

AI as a Teaching Assistant: AI will support teachers by automating administrative tasks, grading assignments, and even identifying areas where students may be struggling. Teachers can then spend more time focusing on providing personalized support to students.

Virtual Classrooms and AI-Driven Tutoring: In a future where education is no longer bound to a physical classroom, young learners might attend virtual classes alongside students from around the world. AI-driven tutoring systems will allow them to ask questions and get instant, personalized feedback.

Continuous Learning: AI will transform education into a lifelong process, with tools and platforms available for learners of all ages. Whether it's formal education, skill-building, or personal enrichment, young learners will have access to learning resources throughout their lives.

Education Focus for AI Integration:

1. STEM Education: Science, technology, engineering, and math (STEM) fields will be crucial for future job markets. Schools should focus on providing students with a strong foundation in these areas, alongside teaching how AI is integrated into these fields.

2. Interdisciplinary Learning: The future workforce will benefit from an interdisciplinary approach to learning. Combining AI knowledge with creativity, ethics, communication, and leadership skills will give students a competitive edge.

3. Online Learning and Flexibility: As AI reshapes industries, learning institutions must also provide flexible, online, and modular education. Platforms like *Coursera* and *edX* allow workers to learn new skills while still employed, ensuring they can upskill without disrupting their careers.

Conclusion: Raising Children in an AI World

The AI revolution brings both excitement and uncertainty for parents like you who want the best for their children. The world young learners will grow up in will be filled with incredible opportunities, as well as challenges. By embracing AI and preparing them with the right mindset, skills, and values, you'll help them not only survive but thrive in this future.

In this new era, it's about equipping the next generation to adapt, learn, and grow alongside AI—not just to use it but to co-create the future with it. Young learners will be part of something truly remarkable, and with your guidance, they'll be ready for whatever comes next.

Chapter 7

AI AND FAMILY DYNAMICS

The AI Revolution
More Family Time

Artificial intelligence will not only impact the workplace and educational systems, but it will also reshape the very fabric of family life. Just like any major technological shift, AI will influence how families interact, make decisions, and navigate everyday life. Whether it's managing household tasks, balancing work-life responsibilities, or preparing future generations to thrive, the role of AI in the family unit will become increasingly important.

AI IN THE HOUSEHOLD: ENHANCING DAILY LIFE

We are already seeing AI assistants such as Amazon Alexa, Google Assistant, and Apple's Siri embedded into households, offering convenience in daily life. However, the future promises even more advanced AI tools that will go beyond simple tasks like answering questions or setting reminders. These tools will manage household dynamics in increasingly sophisticated ways.

Household Management:

AI can help families manage time more efficiently by acting as a "family organizer." Imagine an AI system that tracks the entire family's schedules, coordinating appointments, after-school activities, meals, and even vacations. This system can automatically adjust to changes, send reminders, and help parents balance their work and personal lives without feeling overwhelmed.

Smart Homes:

AI-integrated smart homes are becoming more common, automating things like temperature control, lighting, security systems, and even grocery shopping. AI will allow families to live in homes that "learn" their habits and preferences, optimizing comfort and energy efficiency. For instance, the house might adjust the lighting and heating based on who is home, saving energy when no one is around.

   Concerns: While AI will make life easier in many ways, there are concerns around data privacy and the potential loss of human skills in managing the household. It's essential to maintain a balance between using AI for convenience and ensuring that family members, especially children, are still involved in household responsibilities.

Meal Planning and Nutrition:

AI-powered devices are capable of suggesting personalized meal plans based on each family member's dietary preferences, health needs, and activity levels. It can even order groceries, recommend new recipes, and track nutritional intake to ensure that the family is eating balanced meals.

   Opportunities: Parents could rely on AI to manage grocery lists, meal prep, and even cooking instructions, freeing up time for more meaningful family interactions.

AI Babysitting:

While AI cannot fully replace human caregivers, there are discussions around AI-powered tools designed to monitor young children, alerting parents if their child is in distress or needs assistance. AI baby monitors could track sleep patterns, breathing, and physical activity, offering parents peace of mind.

Challenges: The idea of AI-driven child supervision raises important ethical questions. How much should parents rely on AI to assist in childcare? Should children be interacting with AI babysitters, or should human connections always come first? As a parent, these are critical considerations as AI technologies evolve.

WORK-LIFE BALANCE: NAVIGATING THE AI-ENHANCED HOME OFFICE

The line between work and family life has been increasingly blurred due to remote work trends, and AI will likely play a key role in further blending these aspects of life. AI can assist families in managing the complexities of work-from-home arrangements, allowing parents to balance professional duties with personal responsibilities.

AI as a Productivity Booster: For parents working from home, AI-driven tools can help optimize their workday by scheduling tasks, organizing meetings, and filtering distractions. AI can handle much of the "busy work," allowing professionals to spend more time with their families. Parents can set their working hours, and AI systems can ensure that work does not encroach on family time.

AI-Powered Childcare Solutions: In a future where more parents may be working from home, AI tools could provide personalized learning or entertainment options for children during work hours. This allows parents to focus on their tasks while ensuring that their children are still engaged, safe, and learning.

Opportunities: Parents working from home could use AI tools to ensure their children are still receiving valuable experiences while they attend to professional obligations.

Challenges: While AI may help parents balance work and home life, it's important not to become over-reliant on technology to fill in for human interaction and bonding. Parents need to ensure that technology does not replace the meaningful time spent with their children.

AI AND EMOTIONAL INTELLIGENCE: STRENGTHENING FAMILY BONDS

One of the most important aspects of family dynamics is emotional intelligence—how family members communicate, empathize, and support one another. While AI can handle many practical tasks, it's also beginning to play a role in promoting emotional health and well-being within families.

AI for Mental Health Support: AI-powered tools, such as chatbots designed for emotional support, are becoming more advanced. These tools can provide resources for family members struggling with mental health challenges, offering them exercises, tips, and even suggestions for professional help. AI can monitor emotional states through subtle indicators such as tone of voice, facial expressions, or digital activity, offering a real-time assessment of mental health.

Opportunities: These AI systems could act as an early warning system, flagging potential issues like stress, anxiety, or depression before they escalate. For parents, this could be particularly helpful in monitoring their children's emotional well-being, especially during stressful periods like exams or family transitions.

Challenges: While AI can help with mental health awareness, it is not a substitute for genuine human empathy, care, and connection. Families will need to use these tools as supplements to their emotional intelligence, not as replacements for real conversations and emotional bonds.

Strengthening Communication: AI tools can assist in improving communication within the family. For example, AI-powered apps can offer suggestions on conflict resolution, helping families navigate misunderstandings and differences more effectively. These tools could provide resources for parents to better understand and communicate with their children, especially during difficult developmental stages.

Opportunities: These tools could offer practical solutions for enhancing family relationships, making communication more effective, empathetic, and constructive.

Challenges: Relying too much on AI for communication may lead to a lack of emotional authenticity. Families must remain mindful that the richness of human interaction comes from real emotions, vulnerability, and spontaneity—something AI, despite its advancements, cannot replicate fully.

PREPARING CHILDREN FOR AI IN FAMILY LIFE

As young learners grow up in a world deeply influenced by AI, they will need to understand its role not only in society but also within their family. Teaching them how to interact with AI responsibly, while maintaining their core human values, will be essential.

Teaching Responsible Technology Use: One of the most important lessons you can teach your children is how to responsibly interact with AI tools. This includes understanding that while AI can enhance life in many ways, it is not a replacement for personal effort, human relationships, or critical thinking. Encouraging them to ask questions about the technology they use and to be mindful of its limitations will prepare them for a balanced relationship with AI.

Balancing Screen Time: As AI becomes more integrated into everyday life, managing screen time will become an even bigger challenge for parents. AI-powered devices can easily become a go-to for entertainment, education, or even babysitting. Establishing clear boundaries around the use of AI devices will help young learners develop healthy relationships with technology.

Encouraging Curiosity and Creativity: While AI can automate tasks and provide solutions, human creativity remains a uniquely valuable trait.

Encourage your children to use AI as a tool to enhance their creativity rather than replace it. Whether they are drawing, building, or imagining new worlds, AI can be a co-creator with them, but their originality should always take center stage.

   Opportunities: AI can open new creative avenues for children by offering them tools to experiment with ideas and projects in ways that were never possible before. For instance, AI art generators or music creators can help them visualize and hear their imaginations come to life.

Teaching Empathy and Emotional Intelligence: As AI becomes more involved in family dynamics, it's crucial to teach children the value of empathy, kindness, and emotional intelligence. While AI can help with many practical and organizational tasks, it cannot replace the depth of human emotions. Ensuring that young learners understand the importance of these traits in their relationships will prepare them for a more balanced interaction with AI in their personal lives.

FAMILY TRADITIONS AND AI: PRESERVING WHAT MATTERS

AI might change how we manage households and communicate, but family traditions—the heart of family life—are something that AI should enhance, not replace. Whether it's holiday celebrations, family dinners, or bedtime stories, these traditions provide stability, belonging, and emotional connection in a world of constant technological advancement.

Enhancing, Not Replacing, Family Rituals: AI should be used to enhance family traditions, making them more enjoyable or efficient. For example, AI could help plan family trips, ensuring that everyone's preferences are taken into account, or even recommend new traditions based on cultural or historical interests.

Creating New AI-Enhanced Traditions: Families could start new traditions that incorporate AI in positive ways.

For example, you might use AI to create a yearly family time capsule that documents memories and milestones. Or you could have AI help create custom stories for children at bedtime, weaving their own experiences into the narrative.

Preserving What's Human: Despite all the technological advancements AI brings, maintaining traditions that emphasize human connection, love, and togetherness will be key. Whether it's cooking a meal together without technological assistance or simply talking about the day without relying on AI to mediate, ensuring that family traditions remain centered around real-life interactions is essential.

DRIVERLESS CARS AND THE FUTURE OF FAMILY TRANSPORTATION

The advent of driverless cars is poised to reshape not just how we travel, but how we live as families. While the idea of self-driving vehicles once seemed like science fiction, they are quickly becoming a reality, and their influence will extend far beyond transportation alone. For families, driverless cars will create new opportunities, convenience, and even reshape the concept of personal freedom for all generations. Let's explore how this new AI-driven mode of transport will influence family dynamics.

A SAFER FUTURE ON THE ROADS

One of the most promising aspects of driverless cars is their potential to reduce accidents. Traditional human driving is prone to error—distractions, fatigue, or simply bad judgment. Driverless cars are powered by advanced AI systems that use a combination of sensors, cameras, and algorithms to process information much faster than human drivers ever could. They can make split-second decisions based on real-time data from their surroundings, eliminating many of the risks associated with manual driving.

Opportunities: For families, this means peace of mind. Parents can trust that their children are safer on the road, whether they're traveling together or in a driverless vehicle on their own.

Teenagers, who are often at higher risk of accidents due to inexperience, could especially benefit from the safety improvements offered by driverless cars. Additionally, families with elderly members might find more freedom as AI takes over the wheel, allowing those who can no longer drive to remain mobile and independent.

Challenges: However, with these benefits come some concerns. Driverless cars are still evolving, and there will be a period of adjustment as the technology and infrastructure improve. While AI may outperform human drivers in many ways, technical malfunctions, hacking, and ethical dilemmas (such as how a car decides to act in an unavoidable accident) are all valid considerations that families will need to be aware of as they adopt this technology.

MORE FAMILY TIME AND CONVENIENCE

Imagine a world where you no longer need to focus on driving during long commutes or road trips. With AI handling transportation, family members can use travel time more productively—whether that means relaxing, bonding, or even working. Driverless cars could act as mobile living rooms, allowing parents and children to engage in activities together that they couldn't do if someone had to focus on the road.

Opportunities: Family road trips will become easier and more enjoyable, with everyone able to participate in conversations, games, or even watch movies together during long drives. Parents can use the extra time to help kids with homework, plan family events, or simply unwind without the stress of navigating traffic.

Challenges: The convenience of driverless cars might come with the risk of reduced mindfulness in travel. The journey itself might become so automated that families may feel disconnected from the experience of driving, which, for many, has been a bonding ritual or learning experience. Ensuring that families don't entirely lose the value of time spent together in more traditional travel settings will be important.

ACCESSIBILITY AND FREEDOM FOR ALL AGES

Driverless cars promise a future where mobility is no longer restricted by physical ability or age. For elderly family members or those with disabilities, driverless cars could offer unprecedented independence. No longer will they need to rely on others for transportation—they can simply summon an autonomous vehicle to take them wherever they need to go.

Opportunities: Parents will no longer need to worry about how their aging parents or young children will get from place to place. Autonomous vehicles will create more freedom for the elderly, allowing them to retain their independence while staying safe. It also opens up possibilities for kids to travel to school or extracurricular activities without needing constant parental involvement, offering convenience and flexibility in busy family schedules.

Challenges: With this increased independence, there are concerns about data privacy and security. Families will need to consider how much personal information they are willing to share with these AI systems, as driverless cars collect and analyze large amounts of data to function safely. The trust placed in these systems to protect this information will be crucial.

ENVIRONMENTAL IMPACT AND SUSTAINABILITY

Many driverless car models are being developed with sustainability in mind, often powered by electricity instead of traditional gasoline. This shift could lead to reduced emissions, cleaner air, and a positive impact on the environment—something families are increasingly concerned about as climate change becomes more urgent.

Opportunities: Families that prioritize sustainability will find a significant benefit in transitioning to autonomous, eco-friendly vehicles. As more electric, AI-powered cars hit the roads, their environmental impact will likely lower, benefiting future generations. This is an important factor for families who want to contribute to a greener, more sustainable world.

Challenges: The transition to driverless cars may be slow and expensive initially, with not all families able to afford the latest models or install necessary infrastructure, such as charging stations, at home. Governments and local communities will need to invest in broader adoption strategies to ensure that all families can benefit from this greener future, not just those with higher incomes.

THE FUTURE OF CAR OWNERSHIP

As driverless car technology evolves, families might not even need to own cars at all. The concept of ridesharing will likely expand, with fleets of autonomous cars providing on-demand transportation. Families could subscribe to a service rather than own a vehicle, reducing the cost and hassle of car ownership.

Opportunities: This shift could lead to significant cost savings for families, who would no longer need to pay for car maintenance, insurance, or fuel. Autonomous ride-sharing services could be available whenever needed, eliminating the need for a personal vehicle. Families who live in urban environments with limited parking or those looking to reduce their environmental footprint will especially benefit.

Challenges: Some families might be resistant to giving up car ownership, as the personal vehicle has been a symbol of independence and status for generations. Additionally, families in rural or less connected areas might still need to own a vehicle due to limited access to ride-sharing services. Balancing personal preference with convenience and cost-effectiveness will be key in this transition.

CONCLUSION: AI AND THE FUTURE OF FAMILY

The future of family life will undoubtedly be shaped by AI in ways we are only beginning to understand. From managing household tasks to improving communication, AI will offer countless opportunities to make family life more efficient, connected, and harmonious. However, the core of what makes a family strong—love, empathy, and shared experiences—must remain front and center.

As a parent, preparing your children for this future means guiding them toward a healthy relationship with technology, one where AI is a tool to enhance their lives rather than control it. By fostering creativity, emotional intelligence, and responsibility, you will help them navigate the AI-driven world with confidence, ensuring that they not only adapt to the changes but also thrive in them.

Chapter 8

AI AND HEALTHCARE

The AI Revolution
Revolutionizing Medicine for Families

AI is transforming healthcare in ways that once seemed like science fiction. From personalized treatment plans to faster, more accurate diagnoses, AI is already having a profound impact on how we approach medicine. But for families, this transformation means much more than technological advancements—it's about better health, well-being, and a future where healthcare is more accessible, affordable, and efficient.

Let's explore how AI is changing healthcare for families and what this revolution means for the health of future generations.

PERSONALIZED MEDICINE FOR EVERY FAMILY MEMBER

One of the most exciting developments in AI-driven healthcare is personalized medicine. AI can analyze an individual's genetic makeup, lifestyle, and environmental factors to create highly personalized treatment plans. Instead of a one-size-fits-all approach, AI allows doctors to tailor medical treatments to the specific needs of each family member.

Opportunities: Imagine a future where your child's treatment for asthma or allergies is specifically designed for their unique body and genetic profile.

Personalized medicine could improve outcomes, reduce side effects, and ensure that treatments are more effective for each family member.

AI can also help track family health history, spotting patterns and suggesting preventive care options, ensuring that health issues are caught early before they become serious problems.

Challenges: With personalized medicine comes the challenge of data privacy and security. Families must feel confident that their medical data is secure and won't be misused. Striking a balance between personalization and privacy will be key to ensuring trust in AI-driven healthcare.

EARLY DIAGNOSIS AND PREDICTIVE HEALTH

AI's ability to process vast amounts of data quickly means it can detect diseases earlier than ever before. For families, this could be lifesaving. AI can identify subtle changes in medical data that may signal the early stages of conditions like cancer, heart disease, or diabetes—conditions that are more treatable when caught early.

Opportunities: Families could benefit from AI-powered health monitoring devices that track vital signs, sleep patterns, and other health metrics in real-time. For example, wearable devices equipped with AI could monitor your child's heart rate, oxygen levels, or glucose levels, providing parents with peace of mind and alerting them to any potential health concerns.

In a broader sense, AI can also predict potential health risks based on family history and lifestyle choices, offering preventive care recommendations that could help families avoid future health problems altogether. This could lead to longer, healthier lives for everyone in the family.

Challenges: While the benefits of early diagnosis and predictive health are undeniable, parents may feel overwhelmed by constant health monitoring.

Finding the right balance between embracing AI's insights and maintaining a sense of normalcy in family life will be important.

Additionally, ensuring that families understand the data and can act on it in meaningful ways is crucial for successful adoption.

VIRTUAL HEALTHCARE: CONVENIENT CARE FOR BUSY FAMILIES

The traditional doctor's office visit is evolving. AI-powered virtual healthcare platforms are making it easier for families to access medical care from the comfort of their own homes. This is especially important for busy parents who may struggle to find time for regular check-ups or doctor's visits.

Opportunities: Virtual healthcare offers a convenient solution for families juggling work, school, and extracurricular activities. Parents can consult with doctors via video calls, and AI-powered chatbots can provide quick answers to medical questions, suggest treatments, or even prescribe medication. This can save time and reduce the need for in-person visits to urgent care centers or emergency rooms.

For children, virtual healthcare can also be less intimidating than a traditional doctor's visit, helping to reduce anxiety around medical care. AI systems can monitor health conditions remotely and offer continuous care, allowing families to stay on top of health concerns without needing frequent office visits.

Challenges: While virtual healthcare offers convenience, it may not completely replace the need for in-person visits, especially for more serious health concerns. Parents will need to discern when virtual care is sufficient and when it's necessary to see a doctor face-to-face. Additionally, ensuring that virtual healthcare services are covered by insurance and accessible to all families will be critical to its widespread adoption.

AI AND PEDIATRIC CARE: TAILORED HEALTH SOLUTIONS FOR KIDS

Children's health is often more delicate and requires specialized care. AI is playing a pivotal role in enhancing pediatric care by offering better diagnostic tools, treatment options, and preventive care tailored to the needs of young patients.

Opportunities: AI can assist pediatricians in diagnosing rare conditions that might be missed by traditional methods. For example, AI algorithms can analyze pediatric imaging scans with incredible accuracy, helping to detect issues like developmental abnormalities, fractures, or early signs of illness.

In addition, AI-powered apps and devices can help parents monitor their children's growth and development. These tools can track milestones, detect early signs of developmental delays, and provide personalized advice to ensure that each child is thriving. This could give parents more confidence in managing their children's health, knowing that AI is there to offer expert insights.

Challenges: One of the challenges in pediatric AI is ensuring that technology designed for adults is adapted appropriately for children. AI tools must account for the unique needs and developmental stages of kids. Moreover, parents will need clear guidance on how to interpret AI-driven insights to avoid unnecessary anxiety or confusion.

MENTAL HEALTH: AI'S ROLE IN EMOTIONAL WELL-BEING

Mental health is just as important as physical health, and AI is opening up new possibilities for families to manage stress, anxiety, and emotional well-being. From AI-powered therapy apps to virtual counselors, mental health care is becoming more accessible and less stigmatized, which is crucial for both parents and children.

Opportunities: AI tools can offer support for parents managing the stresses of work-life balance, caregiving, and household responsibilities.

For example, AI-driven mental health apps can offer personalized coping strategies, mindfulness exercises, or even virtual therapy sessions tailored to the individual's needs.

For children and teenagers, AI can help provide early intervention for issues like anxiety, depression, or bullying. Virtual mental health platforms can offer a safe space for young people to express their feelings, talk through their challenges, and receive advice in a non-judgmental environment. AI's ability to analyze emotional cues from language and behavior can also help detect mental health concerns early on, allowing for timely support.

Challenges: While AI can provide valuable mental health support, there's still a need for human connection in therapy and counseling. AI-powered tools should be seen as a complement to traditional mental health care, not a replacement. Additionally, families may have concerns about the privacy and security of their mental health data and ensuring that AI tools adhere to strict confidentiality standards will be essential.

AI in Elderly Care: Supporting Multigenerational Families
For multigenerational families, AI can play a vital role in caring for elderly parents or grandparents. AI-powered devices can assist with medication management, fall detection, and daily monitoring, ensuring that elderly family members receive the care they need while allowing them to maintain independence.

Opportunities: AI systems like smart home assistants and wearable health monitors can help elderly family members manage chronic conditions, reminding them to take medications, tracking vital signs, and even alerting family members in the event of an emergency. AI can also help with mobility, offering solutions like robotic exoskeletons that assist with walking or lifting.

For families who live far from their elderly loved ones, AI-powered caregiving robots or virtual assistants can provide peace of mind, offering companionship, support with daily tasks, and emergency assistance when needed.

This can allow aging family members to live independently for longer while still receiving the care they need.

Challenges: While AI can provide invaluable support for elderly care, some families may feel uneasy about relying on technology for such personal and sensitive tasks. Ensuring that AI is user-friendly for elderly individuals and that it respects their dignity, and autonomy will be key to its success. Families will also need to feel confident that AI is a complement to, rather than a replacement for, human caregiving.

AI'S ROLE IN REDUCING HEALTHCARE COSTS

Healthcare costs are a significant concern for families, and AI has the potential to make healthcare more affordable and efficient. By reducing administrative burdens, streamlining diagnostic processes, and improving the accuracy of treatments, AI can help drive down the costs of medical care.

Opportunities: For families facing high healthcare costs, AI can offer more cost-effective solutions. AI-driven diagnostic tools can reduce the need for expensive tests, while virtual healthcare platforms can cut down on costly office visits. AI's ability to optimize healthcare workflows can also help reduce wait times and improve access to care, making healthcare more efficient and less expensive for families.

Additionally, AI-powered predictive models can help prevent costly medical conditions by catching health issues early and recommending preventive measures, potentially saving families from the financial burden of serious illnesses.

Challenges: As with any new technology, there may be initial costs associated with implementing AI in healthcare. Families may need assurance that these costs will eventually lead to long-term savings. Additionally, it's important to ensure that AI-driven healthcare solutions are accessible to all families, regardless of income level or geographic location.

AI AND HEALTHCARE:
BRINGING HOPE TO PREEXISTING CONDITIONS

AI in healthcare is more than just a tool—it's a revolutionary shift towards enhancing human lives, offering new possibilities for people living with preexisting conditions. Whether it's diabetes, heart disease, or cancer, AI is actively reshaping how we understand, manage, and treat these illnesses. With AI's deep learning algorithms, it's now possible to analyze vast amounts of data in ways that humans simply couldn't, and this is leading to breakthroughs in how we approach medical treatments.

For instance, AI systems are able to sift through millions of medical records to identify patterns that could suggest more personalized treatment plans. If you or a loved one have a preexisting condition, imagine the comfort of knowing that AI is reviewing every possible avenue, looking for overlooked solutions, and ensuring no stone is left unturned.

AI IN CANCER RESEARCH AND TREATMENT: A PERSONAL TOUCH

Take cancer, for example. With conditions like multiple myeloma, there's a constant need for new treatments and ways to slow down progression. AI-powered technologies are already making significant advances in this space. Machine learning algorithms can now study cancer cells more accurately than ever before, identifying genetic mutations and other indicators that help oncologists design more targeted therapies. AI can even predict how different patients will respond to certain treatments, tailoring the approach to each individual.

One breakthrough is in drug discovery. AI is speeding up the process of finding new drugs, shortening what used to take years into mere months. By simulating thousands of scenarios, AI systems can predict which combinations of drugs might be most effective for treating specific cancers. It's possible that in the near future, patients will receive drugs designed specifically for their type of cancer, maximizing the chances of remission or a cure.

In fact, in the area of multiple myeloma, AI has been instrumental in identifying new compounds and therapies. AI systems like IBM Watson have been used to analyze genomic data and offer treatment options that match the genetic profile of a patient's cancer. This sort of precision medicine is not just theory anymore—it's reality.

AI AND HOPE FOR PATIENTS

The healthcare field is one of the most critical areas where AI is showing its transformative power. From disease prevention and diagnosis to personalized treatments, AI is helping doctors and researchers tackle medical challenges in ways we couldn't have imagined just a decade ago. For those living with preexisting conditions, AI doesn't just offer hope—it is actively improving outcomes and opening new doors that can lead to longer, healthier lives.

A NEW CHAPTER IN THE FIGHT AGAINST DISEASE

As AI continues to evolve, the future of healthcare will be increasingly personalized and proactive. While your doctors may give you timelines based on current meadical knowledge, AI's ever-growing capabilities offer new avenues of hope. Who's to say what new treatment might be discovered tomorrow? As more data is collected and analyzed, and as more research is done, AI could unlock new treatments that extend lives in ways we've only dreamed of.

AI is already working behind the scenes—quietly helping doctors, speeding up research, and delivering solutions that offer patients more time, better outcomes, and most importantly, hope. And for anyone living with a preexisting condition, hope is often the most powerful medicine.

EARLY DETECTION AND DIAGNOSIS: A GAME CHANGER

One of AI's most important contributions to healthcare is in early detection. For many diseases, including cancer, early diagnosis can make all the difference. AI algorithms are being trained to analyze medical images—such as CT scans, X-rays, and MRIs—with remarkable accuracy. By comparing images against vast databases, AI systems can detect abnormalities that might be too subtle for the human eye to catch.

For cancers like multiple myeloma, AI can analyze genetic data to identify high-risk patients even before symptoms appear. The earlier a patient is diagnosed, the sooner treatment can begin, leading to better outcomes. In fact, AI systems are now being integrated into routine screenings in some hospitals, helping detect diseases in their earliest stages, often when they are most treatable.

REVOLUTIONIZING TREATMENT: TAILORED THERAPIES

Once a diagnosis is made, AI's role doesn't stop. Its ability to process and analyze massive amounts of data in real time makes it possible to create personalized treatment plans tailored to each individual patient. Gone are the days of a "one-size-fits-all" approach to medicine. Instead, AI leverages insights from a patient's genetic makeup, lifestyle, and even environmental factors to recommend therapies that are most likely to be effective for that specific individual.

In cases like multiple myeloma, treatment options often include chemotherapy, radiation, or stem cell transplants. AI can analyze a patient's specific cancer cells and suggest which drugs or combinations of drugs might have the best chance of success. It's this level of customization that gives AI the power to change lives. Furthermore, AI is contributing to the concept of "drug repurposing." In other words, AI algorithms can scan existing medications used for one condition and determine if they might also be effective in treating another illness.

This is particularly exciting for diseases like cancer, where every new treatment option is a chance for better survival rates.

AI IN IMMUNOTHERAPY AND BEYOND

One of the most promising areas of cancer treatment today is immunotherapy, a method that helps the body's immune system recognize and attack cancer cells. AI is playing a huge role in the development of immunotherapy treatments by helping identify which patients are most likely to respond to specific immunotherapies.

By analyzing a person's immune profile alongside the genetic markers of their cancer, AI is able to suggest new immunotherapy treatments that might not have been previously considered. This is especially exciting for patients with aggressive cancers, as it can lead to therapies that boost the immune system's ability to fight the disease.

Multiple myeloma, in particular, has seen advances in CAR T-cell therapy, where a patient's own immune cells are genetically engineered to target cancer cells. AI is helping refine these therapies, making them more effective and available to more patients. With AI's assistance, immunotherapy is becoming a more precise and potent weapon in the fight against cancer.

AI-DRIVEN CLINICAL TRIALS: SPEEDING UP THE CURE

One of the biggest challenges in medical research is the time it takes to complete clinical trials for new drugs and therapies. AI is already speeding up this process.

Traditionally, it could take years to collect enough data, recruit patients, and analyze results. However, AI can rapidly simulate clinical trials by analyzing patient data from across the globe, finding trends, and identifying potential solutions much faster than human researchers alone.

For example, AI can identify which patients are the best candidates for a trial based on their genetic and health profiles, reducing the guesswork and making trials more efficient. This leads to faster approvals of new treatments and ensures that patients can access the latest innovations as soon as they are available.

THE POWER OF PREDICTIVE ANALYTICS

AI also plays a vital role in predictive analytics—using data to predict health outcomes. This can be especially useful for those managing long-term conditions. For instance, wearable devices and health apps powered by AI can continuously monitor a patient's vitals, sending real-time data to healthcare providers. This allows doctors to track disease progression and adjust treatments as needed, without requiring frequent visits to the hospital.

In cases like multiple myeloma, AI can predict when a relapse might occur or how the cancer is responding to treatment. This kind of foresight gives patients and doctors the ability to make proactive decisions, offering more control over the disease.

AI FOR CHRONIC DISEASE MANAGEMENT

For patients living with chronic conditions, AI-powered tools offer an opportunity for enhanced disease management. From smart pill dispensers that remind you to take medications on time to AI-driven apps that monitor symptoms and send alerts to healthcare providers, managing many conditions can become less overwhelming.

AI helps patients stick to their treatment plans by providing reminders and support when it's most needed. It can track symptoms, identify side effects, and even adjust medication doses to ensure patients receive the most effective care. By reducing human error, AI also reduces the burden on patients and their caregivers, allowing for a better quality of life.

CONCLUSION: A BRIGHT FUTURE AHEAD

AI is revolutionizing healthcare in ways that will benefit families for generations to come. From personalized treatments to early diagnosis and virtual care, AI has the potential to make healthcare more efficient, affordable, and accessible. By embracing AI's potential while addressing the challenges it presents, families can look forward to a future where healthcare is not only more effective but also more attuned to their unique needs.

As we look ahead, the possibilities of AI in healthcare seem boundless. From gene editing techniques like CRISPR that are being advanced with AI's help to the development of entirely new treatments, AI has the potential to dramatically extend lives. Multiple myeloma, once considered incurable, is now a prime focus for AI-driven therapies, and with ongoing research, we may soon see even more breakthroughs.

For countless patients, AI represents a real source of hope. Not just for extending life, but for enhancing the quality of life, providing personalized care, and ultimately, finding cures.

The journey ahead is full of promise, and with AI by our side, we are more empowered than ever to face the future with optimism.

Chapter 9

AI AND EDUCATION

The AI Revolution
Preparing the Next Generation

The world of education is undergoing a radical transformation thanks to AI, which is already reshaping how students learn, how teachers teach, and how schools operate. AI promises to make education more accessible, personalized, and efficient, creating an environment where every student can thrive, no matter their learning style or pace.

For parents, especially those raising young children, the future of education with AI offers exciting opportunities as well as some challenges. It's important to understand how AI will impact not only the classroom but also the entire educational journey from preschool to higher education and beyond.

PERSONALIZED LEARNING FOR EVERY CHILD

One of AI's greatest contributions to education is its ability to create personalized learning experiences. Gone are the days of a rigid, one-size-fits-all approach where every student is taught the same way, at the same pace, regardless of their individual strengths or weaknesses. AI systems can now adapt to the needs of each student, offering lessons and activities tailored to their specific learning style and progress.

Imagine young learners sitting in their classroom, where AI-driven platforms monitor their performance in real-time.

If one is excelling in math but struggling with reading comprehension, the AI will adapt the lessons, accordingly, providing more challenging math problems while offering extra support and practice in reading. This type of personalized learning helps ensure that no child is left behind and that each student can progress at their own pace.

These systems can also provide insights to teachers, alerting them to students who might need extra help or suggesting new strategies for engaging a particular child. As AI continues to advance, it will become an indispensable tool for educators, allowing them to focus more on mentorship and guidance rather than administrative tasks and grading.

AI AS A LIFELONG LEARNING COMPANION

For the children of this generation, AI won't just be a part of their early schooling; it will likely be a lifelong learning companion. As AI becomes more integrated into every facet of life, the ability to continuously learn and adapt will be essential. This is where AI's role in education becomes even more significant.

Learning will no longer be confined to traditional schooling years. AI will enable individuals to pursue knowledge at any stage of life, ensuring that skills are constantly updated in response to changing industries and technologies. For example, adults looking to switch careers or gain new skills in the future will have access to AI-powered platforms that deliver personalized learning modules, assessments, and certifications—all from the comfort of their homes.

This shift toward lifelong learning is particularly important in a world where the job market will be continuously evolving. By embracing AI, we can ensure that future generations, are prepared for whatever challenges and opportunities come their way.

BEYOND THE CLASSROOM: AI AS A LEARNING MENTOR

As AI becomes more integrated into educational environments, it will also expand its role beyond the classroom. Today, we already see AI-powered tutors, chatbots, and apps that help students learn in fun and engaging ways outside of school hours. These tools can provide instant feedback, answer questions, and guide students through difficult concepts, acting as a personal tutor.

For example, let's say a child is struggling with a particular homework assignment. Rather than waiting for help the next day, they can use an AI-powered learning assistant to walk them through the problems, explain the concepts, and offer examples in real-time. This kind of interactive support will be crucial in keeping students engaged and motivated to learn independently.

As AI becomes more sophisticated, it may even evolve into a "learning mentor," helping students explore their interests, set educational goals, and discover new areas of passion. With access to vast libraries of information, AI will be able to suggest books, articles, videos, and interactive experiences tailored to a student's interests, making learning a more dynamic and enjoyable process.

THE ROLE OF TEACHERS IN AN AI-POWERED WORLD

While AI will play a significant role in the future of education, it will not replace teachers—far from it. In fact, AI will likely enhance the role of educators, allowing them to focus more on fostering critical thinking, creativity, and emotional intelligence, areas where human insight is irreplaceable.

AI can take over time-consuming tasks like grading papers, tracking attendance, and even creating lesson plans, freeing up more time for teachers to build meaningful relationships with their students. This will allow teachers to concentrate on what they do best: inspiring curiosity, encouraging growth, and nurturing the potential in each child.

Moreover, teachers will be instrumental in guiding students through the ethical and social implications of AI. As AI becomes more integrated into everyday life, understanding how to use it responsibly and thoughtfully will be essential. Teachers will play a key role in fostering discussions about AI's impact on society, helping students navigate this new world with wisdom and empathy.

AI IN EARLY CHILDHOOD EDUCATION

For younger children, AI can enhance learning in fun, interactive, and age-appropriate ways. Educational apps and games powered by AI are already helping children develop foundational skills in literacy, numeracy, and problem-solving. These tools often use playful animations, engaging characters, and adaptive learning techniques to keep young learners engaged.

For example, AI-powered storybooks can adapt their narratives based on a child's reading level, providing a personalized experience that grows with them. Interactive games can teach early math concepts through puzzles and challenges that adjust in difficulty as the child progresses. These kinds of AI-driven tools can make learning an enjoyable and rewarding experience for young children, laying a strong foundation for future success.

PREPARING CHILDREN FOR AN AI-DRIVEN WORLD

As exciting as AI is, it's also important to prepare children for the challenges and responsibilities that come with it. This involves teaching them not only how to use AI technology but also how to think critically about it. In an AI-driven world, skills like problem-solving, creativity, and ethical reasoning will be more valuable than ever.

Children will need to understand the limitations of AI, how to evaluate its outputs critically, and how to collaborate with AI to achieve their goals.

They will also need to be aware of the ethical implications of AI, such as privacy concerns, bias in algorithms, and the impact of automation on jobs and society.

Teaching children to approach AI with both curiosity and caution will help ensure that they grow up as responsible digital citizens.

As parents, guiding your children through this new landscape will be crucial. It's natural to have concerns about how AI will impact their lives, but by embracing AI as a tool for learning and growth, we can help young learners develop the skills and mindset they need to thrive in the future.

AI'S ROLE IN TOMORROW'S CLASSROOMS

In the future, classrooms may transform beyond our imagination. AI and virtual reality (VR) could revolutionize the way students engage with lessons, making the learning process far more immersive and adaptable to the needs of every child. Here's how this shift may unfold:

PERSONALIZED LEARNING FOR SPECIAL NEEDS STUDENTS

AI has the potential to tailor education to each student's unique needs. For kids who may struggle to keep up with traditional methods, such as slow learners or those with learning disabilities, AI can offer personalized solutions. Imagine an AI-powered tutor that understands each student's strengths and challenges, creating lesson plans and exercises at a pace that's perfectly suited for them. Instead of falling behind, these students would receive real-time feedback, allowing them to master concepts before moving on.

For example, children with dyslexia could use AI tools to help decode words and strengthen reading comprehension. Similarly, AI could support students with ADHD by offering interactive lessons that keep them engaged and focused, rather than relying on one-size-fits-all teaching methods.

VIRTUAL REALITY: BRINGING HISTORY TO LIFE

While we love books and the beauty of reading, we can also embrace the future by blending traditional learning with immersive experiences. Imagine students learning about ancient Egypt not just through textbooks, but by walking through a VR reconstruction of the pyramids. They could witness the construction of the Great Wall of China or watch the signing of the Declaration of Independence unfold in front of them.

These VR settings wouldn't just be limited to history. For science, students could enter a virtual cell to explore its inner workings or even take a spacewalk in a VR version of the International Space Station. These rich experiences would make learning more engaging and help students grasp complex concepts with greater ease.

INCLUSIVITY IN THE CLASSROOM

AI also holds the potential to bridge the gap for students with disabilities, making classrooms more inclusive. For example, visually impaired students could benefit from AI that converts text to audio, or even from VR worlds specifically designed for sensory learning. Children with mobility issues could use VR to experience environments they wouldn't otherwise be able to access physically.

AI-powered communication tools could support students with speech or hearing impairments by providing real-time captions or sign language interpretation during class discussions. By leveling the playing field, AI ensures that every child, regardless of their abilities, has the opportunity to participate and thrive.

BALANCING TRADITION WITH INNOVATION

As much as we embrace new technologies, there will always be a place for books in education. Reading books develops imagination, critical thinking, and the ability to engage with complex ideas in ways that technology might never fully replicate.

However, the key will be in finding harmony between traditional learning tools and AI-driven advancements.

In tomorrow's classrooms, we might see students alternating between reading novels and interacting with VR experiences, both of which nurture creativity and learning in complementary ways. The challenge will be ensuring that we maintain the deep reflective thinking that books inspire while also embracing the dynamic possibilities that AI and VR offer.

Preparing Our Children for the AI Revolution

As a parent, the future of education can seem both exciting and overwhelming. How do we prepare our children for a world where technology changes so rapidly? The answer lies in nurturing a mindset of adaptability and lifelong learning. By teaching our children to embrace AI as a tool that enhances their abilities rather than replaces them, we can help them navigate this evolving world confidently.

Encouraging children to engage with AI from an early age – through interactive learning apps, coding games, or even building their own simple AI models – will not only spark their curiosity but also prepare them for a future where AI is intertwined with their daily lives.

In the future, schools may focus less on rote memorization and more on critical thinking, creativity, and collaboration – skills that are harder for machines to replicate. This shift will encourage children to see AI not as a competitor, but as a tool to amplify their own intelligence.

ADDRESSING CONCERNS AND FOSTERING A LOVE FOR LEARNING

For parents, one of the biggest concerns with AI and VR in education may be the fear of losing the human connection or the worry that children will become too reliant on technology.

It's essential to remember that AI is a tool, not a replacement for teachers or parental guidance. Teachers will still play a pivotal role in guiding students, offering emotional support, and fostering a love for learning.

By using AI to free up time from administrative tasks or individualized lesson planning, teachers can focus more on inspiring students, leading discussions, and encouraging creative exploration. This blend of human touch and technological assistance could create a more nurturing, balanced, and effective learning environment.

AI AND ACCESS TO EDUCATION: BRIDGING THE GAP

One of AI's most transformative promises is its potential to make education more accessible to students around the world. In many countries, children still lack access to quality education due to factors like poverty, geography, or social inequality. AI can help bridge this gap by providing digital classrooms, remote learning opportunities, and language translation tools that make education available to anyone with an internet connection.

AI-powered educational platforms can deliver lessons in multiple languages, offer personalized tutoring, and provide access to digital textbooks and resources. For children in underserved communities, this means they can receive the same quality of education as students in more privileged areas, helping to level the playing field and create more opportunities for success.

In the future, we may even see AI-powered virtual classrooms where students from around the world can come together to learn, collaborate, and share ideas. By democratizing education, AI has the potential to create a more inclusive and equitable global learning environment.

CONCLUSION: PREPARING YOUNG LEARNERS FOR THE FUTURE TOGETHER

As AI continues to evolve, it will play an increasingly important role in education—shaping how we teach, how we learn, and how we prepare for the future. For parents, the challenge is to embrace this technology while also providing your children with the guidance, support, and critical thinking skills they need to thrive in an AI-driven world.

Young learners are growing up in an extraordinary time, where AI will be a natural part of their everyday lives. By preparing them now—teaching them how to learn, how to adapt, and how to think critically—you're giving them the tools they need to navigate this new world with confidence and curiosity.

Chapter 10

AM I TOO OLD?

The AI Revolution
Navigating the AI Revolution at Any Age

One of the most common concerns when new technology emerges is the fear that it's "too late" to learn or that it's only for the younger generation. Many people feel overwhelmed by the rapid pace of change and wonder if they can keep up with a world dominated by AI. But here's the truth: no matter your age, AI is for you, too. In fact, AI can enhance your life in ways you might not have imagined.

THE AGE OF WISDOM MEETS THE AGE OF AI

There's a profound advantage to being older in an AI-driven world: life experience. While younger generations may be more familiar with technology, older individuals possess the wisdom, critical thinking, and decision-making skills honed over years of personal and professional experiences. These qualities are invaluable in navigating this new landscape.

AI tools are not meant to replace human intuition, empathy, or creativity—they are meant to enhance them. Whether you are 50, 60, or even 80, your insights and experiences can be amplified by AI, allowing you to achieve more, both personally and professionally.

LIFELONG LEARNING AND AI: YOU'RE NEVER TOO OLD TO LEARN

Lifelong learning is no longer just a phrase; it's a necessity. Fortunately, AI makes the process of learning easier, more accessible, and more enjoyable. Tools like online courses, AI tutoring systems, and adaptive learning platforms can help anyone learn at their own pace.

These platforms are designed to adjust to your learning style, offering personalized pathways that can help you master even the most complex AI concepts.

Whether it's understanding how AI works or learning to use AI-powered tools, you don't need to be a tech expert. Resources are available to guide you step-by-step, with plenty of support along the way.

ENHANCING YOUR CURRENT SKILLS WITH AI

It's important to remember that AI doesn't require you to start from scratch. No matter what skills you've developed over your lifetime, AI can help you enhance them. If you're a writer, AI can be your research assistant. If you're a teacher, AI can provide personalized lessons for each student. If you're a business professional, AI can analyze market trends and help you make data-driven decisions.

Your existing expertise combined with AI can result in tremendous growth. For example, AI-powered tools like Grammarly can improve your writing. Design software like Canva, enhanced by AI, can take your visual projects to the next level, even if you don't have a background in design. With AI, the possibilities for personal and professional development are boundless.

OVERCOMING THE FEAR OF CHANGE

Fear of the unknown is a natural reaction, especially when it comes to technology. But it's important to remember that AI is here to make life easier, not more complicated.

The key is to start small. Pick one area where AI can assist you—whether it's automating your daily tasks, improving your communication skills, or exploring a hobby—and build your confidence from there.

The misconception that older adults can't adapt to new technologies has been proven wrong time and again. From learning to use smartphones to mastering video calls during the pandemic, older generations have shown resilience and adaptability. AI is no different. It's a tool that can empower you to stay connected, productive, and engaged.

HOW AI CAN ENHANCE YOUR QUALITY OF LIFE

AI isn't just about work or education; it can enhance your personal life, too. For example, AI can help manage health conditions, reminding you to take medication or monitoring vital signs. Virtual assistants like Alexa or Siri can simplify your day-to-day activities, from setting reminders to controlling smart home devices.

AI also offers countless opportunities for staying connected with loved ones, regardless of distance. AI-powered video platforms can translate languages in real-time, so you can connect with friends and family across the globe. For grandparents, AI tools can make storytelling and interactive games with grandchildren more engaging and immersive.

FINDING PURPOSE WITH AI

For many, the later stages of life are a time to reflect on purpose and legacy. AI can play a role here, too. Whether you want to document your life story through a personal blog or create something entirely new, AI can help you share your knowledge, wisdom, and creativity with the world. Writing a book, launching a podcast, or even mentoring younger generations using AI-powered platforms can give you a renewed sense of purpose.

Many older individuals are already using AI to fuel new passions. From AI-assisted painting and music composition to leveraging AI in community-building activities, there are endless ways to stay active and engaged.

THE GREAT EQUALIZER

One of the most beautiful things about AI is that it levels the playing field. It doesn't discriminate based on age, experience, or background. Everyone has the opportunity to benefit from AI's capabilities, no matter where they are in life.

Think of AI as your partner, guiding you through this new world and empowering you to make the most of it. You bring the wisdom and experience, and AI brings the tools and efficiency. Overcoming Age-Related Stereotypes

As AI becomes more integrated into society, there's an opportunity to break down age-related stereotypes. The idea that technology is only for the young is outdated. Many of the most successful leaders, creators, and innovators of our time have been those who continuously adapted, regardless of age.

With the help of AI, you can redefine what it means to be "older" in this digital age. Whether you're mastering new tools, leading projects, or simply staying connected to your passions, you are never too old to embrace AI.

For many older adults, the world of AI and modern technology can feel overwhelming and intimidating. You are not alone—many seniors have similar worries about complex tech systems or AI-driven devices, believing that they might somehow cause harm or skyrocket their bills. But in reality, AI can be an incredibly helpful and accessible tool for them, just as much as for younger generations.

MYTH-BUSTING: AI ISN'T OUT TO GET YOU

One of the first hurdles for many older individuals is overcoming the myths and fears that surround AI. For some, there's an assumption that AI systems are so complex that they'd need to become a tech expert just to use them.

Others may worry that AI might invade their privacy, track them too closely, or cost them financially if they make a mistake. These concerns are understandable, but often exaggerated.

In truth, AI technologies are designed with user-friendliness in mind. Companies developing AI-powered products are working hard to create intuitive, easy-to-use interfaces. And with voice-command systems becoming the norm, interacting with AI can be as simple as asking your phone or smart speaker to play a song, call a loved one, or set a reminder.

ACCESSIBILITY AND AI: MAKING LIFE EASIER

As physical limitations like sight, hearing, and mobility become more common with age, AI is stepping in to fill the gaps. For example, voice-activated devices allow users with poor vision to complete tasks without needing to see the screen. Voice assistants like Siri, Alexa, and Google Assistant can read out messages, emails, and even news articles, making it easy for individuals with visual impairments to stay connected and informed. Screen readers and magnifiers help improve visibility, while audio feedback ensures that no one is left in the dark when interacting with a device.

For those hard of hearing, AI-assisted hearing aids are advancing rapidly. These modern hearing aids do more than amplify sound—they can automatically adjust based on the environment, filter out background noise, and even translate speech into text on a connected device in real-time. This technology offers a lifeline for older adults who may struggle with conversation or miss important details in a noisy room.

Mobility challenges can also be addressed through AI-powered robots and smart home devices. These tools can assist with household chores, cooking, or even simple tasks like turning lights on and off with a voice command. AI systems are designed to provide support without requiring a person to adapt their behavior, empowering seniors to live more independently for longer.

OVERCOMING THE LEARNING CURVE: START SMALL AND BUILD CONFIDENCE

One of the best pieces of advice for seniors who are unsure about AI is to start small. Begin with basic voice-command features on a smartphone or try a simple smart home device like a voice-controlled light or thermostat. Each little success builds confidence and erodes the myth that AI is too complex or scary.

AI tools can also offer companionship. For example, virtual assistants can engage in simple conversations, keep track of daily habits, and even play games with users. These interactions, though simple, can bring joy and a sense of connectedness to older adults who may feel isolated or lonely.

AI-powered devices are designed to be as intuitive as possible, meaning they learn from user interactions. As someone continues using a device, it becomes more personalized to their needs. For instance, an AI-driven health app may start suggesting daily reminders for medication or offer helpful wellness tips without needing any setup by the user. These conveniences make AI a true partner, rather than a hurdle.

REMOVING THE FEAR: AI AS A FRIENDLY COMPANION

The idea that technology is a cold, impersonal force is one of the biggest myths we need to dispel. Many older adults grew up without these tools, so it can feel alien at first. But the truth is, AI is designed to be helpful, efficient, and—most importantly—compassionate in its own way. It is more than capable of adapting to individual needs, helping users feel safe and supported.

For older individuals, there is a growing range of AI-driven health solutions that offer hope. Smart health monitors can track vitals, remind individuals to take their medication, and alert caregivers in case of an emergency. These tools provide an added layer of safety, while giving both seniors and their families peace of mind.

AI is not something to be afraid of; it's here to help. With tools that listen, adapt, and assist, it's becoming more accessible and capable of improving daily life. Older individuals can find comfort in knowing that they can interact with these devices with ease—using them as trusted friends rather than mysterious machines.

GETTING STARTED: AI TOOLS FOR SENIORS

1. *Voice Assistants*: Siri, Alexa, Google Assistant
   - Easy to use and capable of performing tasks like making calls, setting reminders, reading news, or even telling jokes.

2. *AI-Driven Health Apps*:
   - Apps like Medisafe or HealthTap allow seniors to monitor their medication schedules, track symptoms, and consult with healthcare providers remotely.

3. *Hearing Aids and Visual Tools*:
   - AI-powered hearing aids and screen readers offer improved accessibility, making it easier for individuals with sensory impairments to navigate the world.

4. *Smart Home Devices*:
   - AI-driven devices like smart thermostats, lights, and security systems can help seniors maintain independence in their homes, with minimal effort required.

5. *AI Companionship*:
   - Companion robots like ElliQ or social apps powered by AI offer conversation, games, and engagement that can combat loneliness.

Chapter 11

FAMILY LIFE AND AI

The AI Revolution
A New Kind of Connection

We had explored how AI could play a role in improving household management, and we expanded on driverless cars and their impact on family dynamics. Now, let's dive deeper into how AI is transforming the ways families connect, communicate, and grow together.

AI-DRIVEN HOUSEHOLD ASSISTANTS: MORE THAN JUST A GADGET

Many families are already embracing AI through smart home devices such as voice-activated assistants. From controlling household lighting and temperature to ordering groceries, these tools make it easier to manage day-to-day responsibilities. But beyond convenience, AI has the potential to become a central hub for communication, learning, and bonding in families.

Think about an AI assistant that helps plan family schedules, coordinate activities, and even suggests shared experiences like family movie nights or local events. This could alleviate the stress of planning and organizing, allowing more quality time for connection. Imagine a system that grows with the family, understanding preferences and suggesting learning tools for children or resources for parents based on everyone's evolving needs.

ENHANCING EMOTIONAL BONDS WITH AI

We are in a unique era where AI is starting to play a role in emotional wellbeing. Future AI companions may be able to help with emotional regulation and act as virtual counselors. For instance, a family member feeling overwhelmed or stressed could engage with AI-driven mental health apps to receive support in the form of meditation, journaling, or advice. These tools are already in development, but they may become even more integrated into family life in the coming years.

For children, AI tools like emotional learning apps can help teach empathy and emotional intelligence, fostering deeper relationships between siblings and parents. These apps could create a space where children learn to express their feelings more openly and in healthy ways, improving emotional connections within the family unit.

AI IN PARENTING:
PERSONALIZED GUIDANCE FOR CHILD DEVELOPMENT

Imagine personalized AI-based parenting assistants that provide tailored advice for each stage of a child's development. These assistants could analyze sleep patterns, eating habits, and behavioral changes, offering real-time suggestions for improving health and wellness. For instance, AI could detect early signs of developmental delays and recommend specific educational tools to support growth.

Similarly, AI could assist parents by offering insights based on aggregated data from countless other families, helping them make informed decisions about schooling, nutrition, or activities. While the AI won't replace a parent's intuition, it can become an additional resource to support the family's well-being and development.

THE ROLE OF AI IN FAMILY COMMUNICATION

As children grow and families become more tech-integrated, communication within the household may evolve. AI-powered devices can help families stay connected by keeping track of everyone's schedules, providing reminders, and even mediating disagreements. AI systems could suggest conflict-resolution strategies based on each family member's personality and previous interactions. For long-distance families, AI tools like virtual reality and holographic calls could simulate face-to-face conversations, deepening emotional connections despite physical distances.

ADDRESSING CONCERNS:
THE BALANCE OF AI AND HUMAN CONNECTION

While the benefits are promising, it's important to address concerns around too much reliance on AI. Will families lose the ability to communicate directly if AI handles their schedules or resolves disagreements? Will AI become a crutch that diminishes human connections? These are valid questions that we, as a society, must navigate.

A healthy balance is key. AI should be seen as a tool to enhance human relationships, not replace them. Families must remain vigilant about maintaining strong, direct lines of communication, prioritizing human-to-human interaction whenever possible.

AI AND THE FUTURE OF FAMILY RITUALS

We all have rituals that strengthen family bonds—whether it's holiday traditions, weekend breakfasts, or bedtime stories. AI can support these rituals by freeing up time or offering suggestions for new family activities. However, it's crucial that AI doesn't take over these rituals. The magic lies in shared human experiences, and AI's role should be that of an enabler, not the centerpiece of the family dynamic.

Chapter 12

AI AND THE FUTURE OF ART AND CREATIVITY

The AI Revolution
The Pope Dj

While much of the discussion around AI centers on its impact on industries like healthcare, education, and transportation, one of the most profound—and controversial—areas where AI is making waves is in the world of art and creativity. This chapter explores the ways in which AI is reshaping the creative process, its implications for artists and musicians, and the ethical debates that arise from a world where machines can create alongside, and sometimes even in place of, humans.

THE RISE OF AI-GENERATED ART

AI's ability to generate art has grown exponentially in recent years. With machine learning algorithms, AI systems can analyze vast amounts of data, including thousands of pieces of artwork, to learn various artistic styles and techniques. This allows AI to create paintings, drawings, sculptures, and even music that can mimic the work of human artists or invent entirely new styles.

Perhaps the most famous example is the AI-generated portrait titled *Portrait of Edmond de Belamy*, which was created by an algorithm and sold at auction for over $400,000. This landmark event not only stunned the art world but also raised important questions about what constitutes art, creativity, and ownership in an age where machines can produce works that rival human creations.

HOW AI IS CHANGING THE CREATIVE PROCESS

For many artists, AI isn't a threat but a new tool that can be integrated into their creative process. AI can help artists explore new techniques, experiment with different styles, and push the boundaries of what is possible in art. By collaborating with AI, artists can focus on the conceptual and emotional aspects of their work while relying on AI to handle some of the technical elements.

For example, AI can assist artists by generating ideas based on input, creating visual or musical elements that an artist can then build upon. In this way, AI acts as a creative partner, offering suggestions, experimenting with possibilities, and even surprising the artist with unexpected results. This collaboration can inspire entirely new forms of expression that may not have been possible through traditional methods alone.

AI IN MUSIC: A NEW KIND OF COMPOSER

AI's impact on music is just as profound. AI-powered music composition tools are now capable of creating entire pieces of music, from classical compositions to modern electronic beats. These AI systems can analyze vast libraries of music, learning patterns, structures, and melodies to compose new pieces that follow similar rules.

For musicians, this opens up new creative opportunities. Some artists use AI as a collaborative tool, feeding it ideas and allowing it to generate music that they can then refine. AI can also generate background scores for films, video games, or commercials, freeing human composers to focus on more complex or emotionally driven pieces.

The potential of AI in music composition has sparked debates about authorship and originality. Can a piece of music truly be considered art if it's composed by an algorithm? Is the human touch essential to evoke emotion, or can AI successfully mimic that aspect of the creative process?

These are questions that artists, audiences, and legal experts are still grappling with as AI's influence in the music world grows.

ETHICAL AND PHILOSOPHICAL QUESTIONS: CAN AI BE CREATIVE?

One of the most important questions in the discussion of AI and creativity is: Can AI truly be creative, or is it simply following patterns learned from human works? Creativity is often viewed as something inherently human—an expression of emotion, experience, and individuality. AI, by contrast, lacks emotion, consciousness, or personal experience, which raises doubts about whether it can ever create art in the way humans do.

While AI can generate works that are visually or aurally appealing, critics argue that these creations lack the depth and intention of human-made art. AI art, they contend, is more about the data it has been trained on and the algorithms it uses rather than any true creative intent.

However, proponents of AI creativity suggest that art is not solely defined by emotion or intent. Instead, they argue that art can be judged on its impact on the audience, regardless of whether it was created by a human or a machine. If AI-generated art or music moves people, challenges their perceptions, or sparks a dialogue, can it not be considered art in its own right?

AI AND THE DEMOCRATIZATION OF CREATIVITY

One of the most exciting aspects of AI's role in the creative arts is its potential to democratize creativity. Traditionally, creating art or music required specific skills, training, and resources, making it difficult for many people to express themselves artistically. However, AI tools are lowering these barriers, enabling anyone with access to the technology to create.

AI-powered platforms like DALL·E (for visual art) or Amper (for music composition) allow users to input a simple idea or theme and receive a fully realized piece of artwork or a musical composition in return. This accessibility is giving more people the chance to engage in the creative process, regardless of their background or technical abilities.

In this way, AI is expanding the creative landscape, empowering individuals who may not have otherwise had the opportunity to create. Whether these AI-generated works are seen as "true art" is a matter of personal interpretation, but the fact that more people can engage in creative endeavors is a positive step toward broader artistic participation.

AI IN FILM AND LITERATURE: WRITING STORIES, DIRECTING SCENES

AI isn't limited to visual art and music—it's also making strides in film and literature. AI-powered tools can now generate entire short stories, screenplays, or poems based on simple prompts. In some cases, AI has even been used to co-write novels or scripts alongside human authors, providing a new type of creative partnership.

In the world of film, AI can assist with everything from scriptwriting to editing. For example, AI can analyze vast amounts of film footage, selecting the best takes or suggesting edits that enhance the storytelling. Directors and editors can then refine the work, using AI's suggestions as a foundation.

AI's involvement in literature and film raises many of thea same questions that surround AI in music and visual arts. What role does human emotion and experience play in storytelling? Can an AI-generated story ever truly capture the complexity of human life, relationships, or struggles?

As AI becomes more integrated into the world of film and literature, these questions will continue to shape discussions about the future of art and creativity.

AI AND COPYRIGHT:
LEGAL QUESTIONS IN THE AGE OF CREATIVITY

As AI continues to redefine art and creative industries, a new set of legal and ethical challenges around copyrights has emerged. Artists, musicians, writers, and creators are concerned about the implications of AI-generated content, especially in terms of ownership and intellectual property. Understanding these legalities is crucial for anyone navigating the intersection of AI and creativity.

COPYRIGHT AND AI: WHO OWNS THE ART?

One of the most debated topics is the ownership of AI-generated works. Traditional copyright law is designed to protect works created by humans, but AI blurs this line. If an algorithm creates a piece of music, a digital artwork, or even a book, who holds the rights to that work? The current stance in most legal systems is that AI, as a non-human entity, cannot hold copyright. This means the creator of the AI (such as the developer) or the person who uses the AI to generate the art may own the rights. However, this can vary based on jurisdiction, and legal frameworks are evolving.

TRAINING MODELS: FAIR USE OR INFRINGEMENT?

A major concern for artists is the process through which AI is trained. AI models are often trained on vast datasets of pre-existing works, many of which are protected by copyright. For example, an AI might learn to generate new paintings by analyzing thousands of existing pieces of art. But is this legal? The legality of using copyrighted works to train AI models falls into a gray area. In some cases, it may be considered "fair use," especially if the AI-generated output is transformative or significantly different from the original content. However, this is still a matter of debate, and some artists argue that training models on their works without permission is a form of exploitation.

THE QUESTION OF ATTRIBUTION

Another issue revolves around attribution. Many artists fear that their work could be used by AI models without proper credit, leading to a loss of recognition and potential income. While some AI-generated works are clearly original, others may closely mimic the style of specific artists. This raises ethical questions: should the original artist receive credit or compensation for their influence on the AI's output?

EVOLVING LEGAL FRAMEWORKS

As AI continues to disrupt creative industries, governments and legal bodies are beginning to explore how to regulate these new technologies. Some countries are considering creating new copyright frameworks that specifically address AI-generated works. Until then, artists should be proactive about protecting their rights, understanding licensing agreements, and advocating for clearer regulations.

WHAT ARTISTS CAN DO

*Licensing Agreements: Artists should consider using licensing terms that restrict how their work can be used in AI training models. Creative Commons licenses, for example, can offer different levels of protection.

*Legal Advice: As the laws evolve, seeking legal advice can help artists navigate these complexities and ensure that their rights are protected.

*Engagement in Policy Discussions: Artists can also participate in discussions around policy-making to help shape the future of copyright law in the context of AI.

THE FUTURE OF ART AND CREATIVITY IN AN AI-DRIVEN WORLD

As AI continues to evolve, its influence on the arts will only grow. While some fear that AI could replace human artists, musicians, and writers, the reality is more likely to be one of collaboration and augmentation. AI will serve as a tool that artists can use to explore new creative frontiers, pushing the boundaries of what's possible and opening up new opportunities for artistic expression.

For those who are concerned about the human element being lost, it's important to remember that AI is not a replacement for human creativity but a complement to it. AI cannot replicate the personal experiences, emotions, and individuality that humans bring to their creations. Instead, it can enhance the creative process, offering new tools, perspectives, and possibilities.

The future of art and creativity in an AI-driven world is one of collaboration, experimentation, and expansion. As artists, musicians, and writers continue to explore the potential of AI, we can expect to see entirely new forms of creative expression emerge, blending the best of human ingenuity with the power of advanced technology.

Chapter 13

PREPARING FOR THE AI REVOLUTION

The AI Revolution
Skills and Adaptability in a Rapidly Changing World

As AI continues to transform industries, the skills that individuals will need to succeed are also evolving. It's no longer just about technical expertise—though that is important—but also about adaptability, creativity, emotional intelligence, and a willingness to continuously learn. Preparing for this AI revolution means embracing new mindsets, developing new skills, and fostering resilience in the face of change.

THE IMPORTANCE OF LIFELONG LEARNING

One of the key aspects of thriving in the age of AI is adopting a mindset of lifelong learning. As AI tools evolve and become more integrated into everyday life and work, people will need to constantly update their knowledge and skills to stay relevant. Whether through formal education, online courses, or self-driven learning, staying adaptable and open to new ideas will be essential for success.

Traditional education systems are already shifting to reflect this need for continuous learning. Online platforms like Coursera, Udemy, and Khan Academy are providing affordable and accessible ways to gain new skills, while companies are investing in upskilling and reskilling their employees to prepare them for AI-driven changes in their industries. For individuals, the ability to learn and unlearn quickly will become a valuable asset.

CREATIVITY AND CRITICAL THINKING: THE HUMAN ADVANTAGE

While AI is capable of automating routine tasks and even generating creative content, human creativity and critical thinking remain irreplaceable. AI can analyze vast amounts of data and recognize patterns, but it lacks the intuition and emotional depth that humans bring to problem-solving and innovation.

In the workplace, employees who can think creatively and approach problems from unique perspectives will continue to thrive. These skills—often referred to as "soft skills"—include empathy, communication, collaboration, and adaptability. As AI takes over repetitive tasks, humans will be free to focus on higher-level thinking, using AI as a tool to enhance their work rather than replace it.

EMOTIONAL INTELLIGENCE AND THE HUMAN TOUCH

In a world where AI handles more technical and analytical tasks, emotional intelligence will become an even more important skill. Emotional intelligence involves understanding one's own emotions and those of others, allowing for better communication, collaboration, and leadership.

As AI systems are increasingly used in customer service, healthcare, and education, the human touch will remain crucial. People still crave empathy, understanding, and human connection, especially in fields where emotions play a large role, like counseling, teaching, and caregiving. Workers with high emotional intelligence will be able to complement AI by offering the interpersonal skills that machines cannot replicate.

ADAPTABILITY: THRIVING IN UNCERTAINTY

The rapid pace of AI development means that the future of work will be more uncertain than ever. Some jobs will disappear, others will be created, and many will be transformed.

The ability to adapt to these changes will be a key factor in determining success in this new world.

Being adaptable means more than just learning new skills. It involves a willingness to embrace change, take risks, and be comfortable with ambiguity. For businesses, fostering a culture of adaptability will be essential for survival, as organizations will need to pivot quickly in response to technological advancements and shifting market demands.

COLLABORATIVE SKILLS: WORKING ALONGSIDE AI

As AI becomes a more integrated part of the workplace, individuals will need to learn how to collaborate effectively with these tools. This means understanding AI's capabilities and limitations, knowing when to rely on AI for data analysis or task automation, and when to trust human judgment and creativity.

In many industries, employees will work in tandem with AI systems, using them to augment their work rather than replace it. For example, in healthcare, doctors may use AI to assist with diagnosing patients, but they will still rely on their own expertise to make final treatment decisions. In marketing, AI can analyze consumer behavior, but creative professionals will still be needed to craft compelling campaigns.

LEARNING AND APPLYING AI IN EVERYDAY LIFE

Natural Language Processing (NLP) Tools
**Why it's important: NLP is everywhere! From chatbots to language translation, AI is becoming part of how we communicate and interact online. It's also one of the most accessible areas of AI for beginners because it revolves around language, something we're all familiar with.

**How to start: Many tools require no coding experience at all, like OpenAI's ChatGPT, Google's Dialogflow for creating chatbots, and various speech-to-text or text-to-speech apps.

Starting here allows users to see real-world applications of AI in action.

AI-POWERED CREATIVE TOOLS

*Why it's important: For artists, writers, and musicians, AI tools designed for creativity can revolutionize the way we create content. They help generate new ideas, enhance artwork, and even produce music or writing drafts.

*How to start: No-code platforms like **RunwayML**, **Canva's AI features**, and **DALL-E** are perfect for creative people looking to explore how AI can assist with graphic design, video editing, and generating images or ideas for their projects.

AUTOMATION TOOLS (AI FOR EFFICIENCY)

*Why it's important: AI can take over repetitive tasks, saving you time and effort, which can be particularly beneficial for those who work in businesses, offices, or freelancing.

*How to start: AI automation tools like *Zapier*, *IFTTT*, and *Microsoft Power Automate* are excellent for automating workflows without needing any coding knowledge. These platforms allow users to connect apps and services (e.g., sending automatic email replies or backing up files to the cloud), empowering people to work smarter, not harder.

Chapter 14

AI ETHICS AND RESPONSIBILITY

The AI Revolution
With Great Power Comes Great Responsibility

A s we explore the vast potentials of AI and the many ways it can transform our lives, it's important to address the ethical and responsible use of AI. With great power comes great responsibility, and while AI offers incredible advancements, it also raises significant concerns about privacy, bias, transparency, and the overall impact on society. This chapter will dive into these issues and how we, as individuals, corporations, and governments, can ensure that AI is used for the greater good.

PRIVACY CONCERNS

AI relies heavily on data. It uses data to learn, improve, and make decisions. However, this leads to a fundamental question: how is our personal data being used, and are we giving away too much information without realizing it?

Data Collection and Consent: Many AI applications collect vast amounts of data from users. For example, voice assistants, social media platforms, and smart devices often rely on user data to function effectively. It's crucial for users to understand how their data is being collected and for companies to offer transparency around how that data is being used.

Security of Personal Information: Data breaches and unauthorized access to personal information are growing concerns.

AI can enhance cybersecurity by identifying potential threats, but it can also be a target for cybercriminals. Users must be vigilant about their data and take steps to protect themselves online.

BIAS IN AI

AI is only as good as the data it's trained on, and unfortunately, data often reflects the biases of the real world. If AI models are trained on biased or incomplete datasets, they may unintentionally perpetuate stereotypes or exclude certain groups.

Examples of Bias: From facial recognition technology that struggles to identify people of color accurately, to AI hiring systems that favor certain demographics, bias in AI can have serious consequences. These examples show how algorithms can unintentionally harm certain populations, even if not designed with malicious intent.

Combating Bias: To reduce bias in AI systems, developers need to ensure diverse and inclusive datasets. It's also essential to regularly audit AI systems and include diverse teams in the development process. Ethical guidelines should be implemented to ensure fairness in AI's decision-making processes.

TRANSPARENCY AND ACCOUNTABILITY

As AI systems become more complex, it becomes harder to understand how they reach certain decisions. This lack of transparency, often referred to as the "black box" problem, can lead to mistrust.

Explainability: It's important for AI systems to be able to explain their decisions in ways that are understandable to humans. For example, if an AI model denies someone a loan, the applicant should have the right to know why that decision was made.

Accountability: When an AI system makes a mistake, who is responsible?

Should it be the developers, the company that deployed the AI, or the AI itself? These are questions society needs to address as AI becomes more integrated into critical areas like healthcare, finance, and criminal justice.

ETHICAL USE OF AI

AI can be a powerful tool for good, but in the wrong hands, it can also be used for harm. The development of autonomous weapons, deepfake technology, and AI-driven surveillance has raised concerns about the ethical use of AI.

Regulating AI: Governments around the world are beginning to take steps to regulate AI to ensure its ethical use. The European Union, for example, has proposed the *Artificial Intelligence Act*, which seeks to impose strict regulations on high-risk AI applications. However, there is still much work to be done to establish global standards and guidelines.

Corporate Responsibility: Companies that develop AI systems must take responsibility for ensuring their products are used ethically. This includes establishing ethical guidelines, being transparent with users, and implementing safeguards to prevent misuse.

AI FOR SOCIAL GOOD

Despite these concerns, AI also offers incredible potential to address some of the world's most pressing challenges. When developed and deployed responsibly, AI can be a force for positive change.

Healthcare: AI is already being used to develop life-saving treatments, diagnose diseases, and improve patient outcomes. With further advancements, AI has the potential to revolutionize healthcare access and quality on a global scale.

Environmental Sustainability: AI can play a critical role in addressing environmental challenges. From optimizing energy usage to tracking deforestation and predicting natural disasters, AI offers innovative solutions to protect our planet.

Education: As we've discussed, AI has the potential to revolutionize education by offering personalized learning experiences and breaking down barriers to access. It can help bridge the gap between privileged and underprivileged students, providing equal opportunities for all.

GLOBAL COLLABORATION

The ethical challenges of AI aren't limited to one country or region. They are global issues that require cooperation between governments, organizations, and individuals. The development of international regulations and ethical standards is crucial to ensure that AI benefits humanity as a whole.

International Efforts: Organizations like the United Nations and the *OECD* have already started discussions on AI ethics. The *Asilomar AI Principles*, for example, are a set of guidelines created by AI researchers and ethicists to promote responsible AI development.

Cross-Sector Collaboration: It's not just governments that need to be involved in AI ethics. Collaboration between academia, industry, and civil society is essential for creating a shared vision of how AI should be used in the future.

CONCLUSION: LAWS BEGIN AS IDEAS

As we embrace the AI revolution, we must also be mindful of its ethical implications. By addressing privacy concerns, combating bias, ensuring transparency, and promoting ethical use, we can harness the full potential of AI for the benefit of all. In the next chapter, we will explore how individuals can prepare themselves and their families for the AI-powered future.

Chapter 15

PREPARING FOR THE AI-POWERED FUTURE

The AI Revolution
Change On The Horizon

As AI continues to evolve, the question isn't whether AI will affect our lives—it already is. The real question is how we can best prepare ourselves, our families, and society for the rapidly changing world that AI is shaping. This chapter focuses on practical steps individuals can take to adapt to AI advancements, with a particular focus on building new skills, understanding AI's impact on work and family life, and fostering a mindset of continuous learning.

LIFELONG LEARNING: EMBRACE THE AI MINDSET

One of the most important aspects of preparing for the AI-powered future is adopting a mindset of continuous learning. In a world where AI is automating many routine tasks, the value of human creativity, critical thinking, and problem-solving is higher than ever. Learning doesn't stop after formal education; it's an ongoing journey.

Stay Curious: AI is moving quickly, and new innovations are introduced regularly. Cultivating a sense of curiosity about emerging technologies will help individuals stay ahead. Whether through online courses, reading books, or attending conferences, there are countless ways to stay informed.

Adaptability: In the future of work, adaptability will be a key asset. The ability to pivot, learn new skills, and embrace new ways of working will be critical for success in an AI-driven world.

Growth Mindset: Individuals with a growth mindset, who believe in their ability to learn and improve, are more likely to thrive in the face of AI advancements. AI might change job roles, but those who are willing to upskill and embrace new opportunities will excel.

KEY SKILLS FOR THE AI ERA

There's no need to be a coding expert to benefit from AI, but there are specific skills that can make individuals more competitive in the future workforce. While technical skills are essential, many human-centric skills will remain irreplaceable and highly valuable.

Data Literacy: Understanding how data is collected, analyzed, and used by AI systems is becoming a foundational skill across industries. Data literacy doesn't require deep technical expertise but an understanding of how AI processes information and draws insights.

Critical Thinking: AI excels at pattern recognition and prediction, but it's still humans who need to interpret AI's results and make complex decisions. Critical thinking will remain an essential skill in evaluating AI's outputs and assessing when human judgment is needed.

Emotional Intelligence: As AI takes over repetitive, logical tasks, uniquely human traits like empathy, communication, and collaboration will become even more important. Jobs that rely on emotional intelligence, such as leadership roles or caregiving positions, are less likely to be replaced by AI.

PREPARING YOUR CHILDREN FOR AN AI FUTURE

For parents, one of the most pressing concerns is how to prepare children for a world that will look vastly different from the one they grew up in. In the AI era, education will play a key role in equipping the next generation with the skills and mindset to thrive.

Encourage Curiosity and Creativity: AI can handle many tasks, but creativity is a uniquely human trait that is hard to replicate. Encouraging children to explore their interests, ask questions, and experiment with ideas will help them cultivate the creativity needed for the future.

STEM Education: While not every child needs to become a computer scientist, understanding the basics of science, technology, engineering, and math (STEM) will provide a strong foundation for their future careers. These skills will be relevant in nearly every industry as AI continues to advance.

Ethics and Responsibility: Teaching children about the ethical considerations surrounding AI is also essential. Children should understand the impact of AI on privacy, fairness, and society. Fostering discussions about responsible technology use will help them become conscientious digital citizens.

THE CHANGING NATURE OF WORK

AI is reshaping the job market, and many industries are already feeling the effects of automation and AI-driven processes. While some jobs may be displaced by AI, others will be created, and new roles will emerge that we haven't yet imagined.

Job Transformation: Rather than replacing all jobs, AI will likely transform many existing roles by automating repetitive tasks and augmenting human capabilities. Jobs in healthcare, education, and customer service, for example, may benefit from AI support, allowing professionals to focus on higher-level responsibilities.

Reskilling and Upskilling: To remain competitive, workers may need to reskill or upskill. This doesn't necessarily mean learning to code, but it does mean acquiring skills that complement AI, such as project management, data analysis, or digital marketing.

Hybrid Work Environments: AI will enable more flexible work environments, with remote work becoming even more common. AI tools can assist in managing distributed teams, tracking productivity, and enhancing communication, creating a more seamless balance between work and home life.

AI FOR FAMILY LIFE

AI won't just change the way we work; it will also transform our personal and family lives. From AI assistants that help with daily tasks to smart home technologies that enhance comfort and security, AI has the potential to simplify and enrich family life.

Smart Assistants: Voice-activated assistants like Siri, Alexa, and Google Assistant can help with everything from setting reminders to managing household tasks. As these tools become more sophisticated, they will offer even greater assistance, from helping with grocery shopping to providing personalized health advice.

AI-Enhanced Learning for Kids: As discussed earlier, AI can play a crucial role in education, particularly in personalized learning. Tools like intelligent tutoring systems can help children learn at their own pace and offer additional support for those who need it.

Family Safety and Security: AI-powered security systems, such as smart cameras and sensors, can enhance the safety of homes and families. These technologies can monitor for potential threats, alert homeowners to unusual activity, and even recognize family members through facial recognition.

FOSTERING DIGITAL LITERACY

In the AI era, digital literacy will be essential for both adults and children. Knowing how to navigate and understand AI-driven systems will become a basic requirement, much like traditional literacy today.

Understanding AI: It's important to understand what AI can and can't do. Many people have misconceptions about AI, thinking it's either a magical solution or a dystopian force. Educating yourself about the capabilities and limitations of AI will help dispel myths and enable informed decision-making.

AI in Everyday Life: AI is already embedded in many aspects of daily life, from social media algorithms to recommendation engines. Being aware of how AI influences the content we see and the decisions we make can help individuals become more critical consumers of technology.

Privacy Awareness: As AI collects data from various sources, individuals must become more mindful of their digital footprint. Understanding how to protect personal data, manage privacy settings, and make informed choices about the services we use is essential.

BUILDING AI SKILLS FOR THE WORKPLACE

For those looking to incorporate AI into their professional life, there are several ways to get started. You don't need a degree in computer science to leverage AI tools, but understanding how to work with AI will give you a competitive edge.

AI Tools for Non-Coders: There are many user-friendly AI tools that don't require coding experience. Platforms like *Google Cloud's AutoML*, *IBM Watson*, and *Microsoft Azure* offer AI services that can be applied to various business needs, from customer service automation to data analysis.

AI in Business Operations: Understanding how AI can optimize operations, such as supply chain management or customer relations, is crucial for professionals in many industries. AI can also help with strategic decision-making by offering predictive insights.

Collaborating with AI: In the future, humans and AI will increasingly work together. Developing the ability to collaborate with AI systems—by providing feedback, overseeing AI decisions, and managing AI-driven projects—will be a valuable skill in any workplace.

AI is set to reshape the workforce, making it crucial for both job seekers and current professionals to build AI-related skills. Even for those without coding experience, familiarity with AI concepts can open doors. Online platforms like Coursera, edX, and Udemy offer AI-focused courses that introduce concepts like machine learning, data analysis, and natural language processing. Basic AI tools for business—such as Microsoft's AI features or Google's AI integrations—are also easy entry points for upskilling. Embracing AI early on allows individuals to future-proof their careers in a rapidly evolving landscape.

Key soft skills like critical thinking, problem-solving, and creativity will remain invaluable as automation continues. Even as AI automates routine tasks, these uniquely human traits will differentiate individuals in the AI-driven workforce. Understanding how to complement AI tools rather than compete with them is the key to thriving in tomorrow's workplace.

The AI revolution is an exciting and transformative time for all of us. By embracing lifelong learning, preparing our children, and developing key skills, we can thrive in this new era. In the next chapter, we'll explore how different industries are being reshaped by AI and the new opportunities that are emerging in various sectors.

Chapter 16

AI AND THE FUTURE OF FAMILY LIFE

The AI Revolution
Enhancing Home and Relationships

As AI enters our homes, its role in family life will go beyond convenience—it will redefine how families interact, share experiences, and maintain balance. While technology has always been present in our personal lives, AI brings a new level of personalization and automation. From virtual assistants to intelligent appliances, AI will influence every aspect of family dynamics, making homes smarter, and perhaps, more harmonious.

AI-POWERED HOMES: THE RISE OF SMART LIVING

The home of the future will be equipped with AI tools that anticipate our needs before we even articulate them. From automated lighting that adapts to mood and time of day, to AI-powered kitchens that suggest recipes based on what's available in the fridge, AI will make life easier and more efficient. AI-enabled cleaning robots, like vacuum cleaners or lawnmowers, will free up time, allowing families to spend more moments together.

AI will also offer safety advantages. Intelligent home security systems will monitor not just for intruders but for accidents, like fires or falls. For families with aging members, this could be a game-changer, providing a sense of security and independence to the elderly, while giving peace of mind to loved ones.

STRENGTHENING BONDS: AI AS A FAMILY MEDIATOR

AI could potentially help in maintaining harmony in family relationships. Intelligent devices will track family schedules, sending reminders about important family events or syncing calendars to ensure that precious moments like birthdays, school performances, or family dinners are never missed. AI will help smooth over the daily logistical challenges that can create stress between family members.

Moreover, AI-based wellness apps could encourage healthy family activities. From suggesting group workouts or meditation sessions to helping organize family trips, AI will not only offer convenience but encourage stronger family ties. By taking care of the mundane, it will leave room for what truly matters—building relationships.

PARENTING IN THE AI AGE: A NEW PARADIGM OF SUPPORT

AI will also revolutionize parenting. Imagine AI-powered baby monitors that can alert parents not only when their child is awake but when they sense potential health issues, such as irregular breathing or elevated temperatures. AI can track developmental milestones and offer suggestions to parents on the best ways to foster their child's growth.

Parents will also have AI as an educational ally. Personalized learning apps tailored to a child's interests and pace will help nurture a love of learning from an early age. These apps will adapt in real-time, providing just the right challenge to keep children engaged and developing at their own unique speed.

Of course, these tools will need to be used in balance. Parents will need to maintain an active role to ensure that AI enhances their children's experiences rather than replacing the need for direct human connection.

THE EVOLUTION OF FAMILY ENTERTAINMENT

Entertainment is another area where AI will make a profound impact. AI-driven virtual reality (VR) environments will allow families to experience immersive adventures together, such as exploring the deep sea or walking through historical periods as though they were actually there. AI will recommend personalized entertainment options—movies, music, games—that cater to the preferences of each family member, ensuring that everyone enjoys what they do together.

Families will also have more options to co-create content. AI-powered video editing software will allow parents and children to easily make home movies, or even animated films. By allowing everyone to be part of the creative process, AI will open up new avenues for collaborative family fun.

THE CHALLENGE: BALANCING TECH WITH HUMAN CONNECTION

As exciting as these advancements are, there will also be challenges. One of the most significant is ensuring that AI doesn't drive families apart by making everyone more individually focused. The very tools designed to bring families together could inadvertently isolate individuals if not used mindfully. Family dinners, conversations, and the "analog" moments of life should not disappear in favor of AI-driven interaction.

Families will need to strike a balance, ensuring that AI enhances life rather than substitutes it. Teaching children about the responsible use of technology will be essential for future generations growing up with AI. This includes establishing boundaries, encouraging face-to-face interaction, and making sure that technology does not take precedence over family bonds.

CARING FOR LOVED ONES: AI AND ELDER CARE

For families caring for older relatives, AI could be a transformative tool. Intelligent home monitoring systems can assist older family members with tasks such as medication reminders, fall detection, and even companionship. AI-powered wearables and smart devices can monitor vital signs in real time, sending updates directly to caregivers.

With AI, family members living far away from their aging loved ones will have peace of mind, knowing they can receive real-time information on their health and well-being. AI companions may also help mitigate the loneliness experienced by many elderly people, offering conversation, reminders, and even entertainment.

While AI will never replace the warmth and love that family members bring, it can supplement the care provided, offering additional support to ensure older individuals maintain their independence while staying connected to their families.

CONCLUSION: THE FUTURE OF FAMILY LIFE IN AN AI WORLD

AI will redefine family life in profound and positive ways. From making daily tasks more efficient to deepening family connections, its presence will be felt in every corner of the home. However, it will be up to families to navigate these changes thoughtfully, ensuring that AI serves to strengthen rather than weaken the human bonds that are central to family life.

The future of AI in homes is exciting, full of possibilities to enhance our living experience. Yet it is also a reminder that technology must be handled with care and intention, especially when it comes to the relationships we hold dearest.

Chapter 17

INDIVIDUAL GOALS

The AI Revolution
Aligning AI with Personal Ambitions

The rapid rise of artificial intelligence presents a wealth of opportunities, but understanding how AI aligns with your personal ambitions is key to reaping its benefits. AI is a versatile tool that can empower anyone to achieve their dreams—whether you're an artist, entrepreneur, educator, or someone looking for personal growth. This chapter will delve into how you can harness AI to achieve your individual goals and why personalizing your AI experience is essential in today's ever-changing world.

REFLECTING ON YOUR PERSONAL VISION

To align AI with your personal goals, you must start by deeply reflecting on your life ambitions. Think about where you are now, where you want to go, and how AI can facilitate that journey. Whether it's a professional aspiration, a creative endeavor, or personal development, having a clear vision is crucial. AI offers many pathways: automation can save time, AI tools can enhance creativity, and intelligent systems can open doors to new knowledge and opportunities.

For instance, an artist might explore AI tools to experiment with new mediums and forms of expression. A small business owner could use AI-powered analytics to boost sales or streamline operations.

Even on a personal level, AI-powered wellness apps could help you better manage stress and mental well-being. The key here is to map AI's capabilities onto your goals rather than adjusting your ambitions to fit the technology.

Action step: Take a few moments to write down three major life goals. For each, note how AI might make achieving these goals faster, easier, or more efficient.

LEARNING AI THROUGH THE LENS OF YOUR INDUSTRY

Every industry has a different relationship with AI, so learning how it applies to your specific field is one of the most efficient ways to approach AI education. Many professions are being transformed by AI, such as healthcare, finance, art, education, and even retail. In healthcare, for example, AI is being used for diagnostics, personalized treatment plans, and predictive health analytics. In finance, AI is increasingly being employed for algorithmic trading, fraud detection, and customer service automation.

If you're a professional, learning about AI's applications in your industry can help you stay ahead of the curve. For instance, marketers should become familiar with AI-driven consumer insights tools, while educators should explore AI platforms that personalize learning for students. Staying informed and adapting to these trends will not only make you more effective but also give you a competitive edge.

Action step: Research the top three AI applications currently being used in your industry. Identify one tool or technology that you can explore and apply to your work.

DEVELOPING A PERSONAL AI SKILL SET

Once you've identified how AI can support your individual goals, it's time to cultivate a relevant skill set. Even if you don't plan to become a full-time AI expert, understanding the fundamentals of AI will empower you to use the technology more effectively.

This includes learning how to work with AI tools, interpreting AI-driven insights, and staying abreast of new developments in the field.

There are countless resources available for learning about AI, from free online courses to hands-on tools that require little to no coding experience. For example, platforms like *Teachable Machine* allow you to create simple AI models without writing any code, while *Udemy* and *Coursera* offer in-depth courses on machine learning, AI ethics, and AI for business.

Building a foundational understanding of AI doesn't have to be overwhelming. Start small, learning only what is immediately relevant to your goals. As you become more comfortable with AI, you can expand your knowledge into more complex areas.

Action step: Commit to learning one new AI skill each month. Start with beginner-friendly platforms, then gradually move on to more advanced topics once you're comfortable.

INTEGRATING AI INTO DAILY LIFE AND PERSONAL PROJECTS

One of the most practical ways to ensure AI aligns with your ambitions is to incorporate it into your daily life and personal projects. This doesn't necessarily mean you have to undertake an AI-heavy project right away. Instead, start by identifying everyday tasks that AI can help with. Whether it's automating your email, using AI-powered design tools, or even applying AI to monitor your fitness routines, there are countless ways to enhance your productivity and creativity with AI.

For example, if you're a writer, tools like *Grammarly* or *Copy.ai* can help refine your writing or generate new content ideas. For those in sales or marketing, AI platforms such as *HubSpot* offer advanced data analytics to drive better decisions. Small improvements in your workflow, multiplied over time, can have a profound impact on both your professional and personal life.

Additionally, larger personal projects such as launching a blog, starting a side business, or even learning new hobbies can be enhanced by AI. Automation tools like *IFTTT* (If This Then That) can simplify tasks, while generative design tools such as *RunwayML* can introduce new creative possibilities. AI should become your silent partner, helping you accomplish more with less effort.

Action step: Choose one personal project you'd like to enhance with AI. This could be anything from designing a new website to improving your health habits. Find at least one AI tool that can help you achieve this goal faster and more efficiently.

ADAPTING AND ADJUSTING YOUR GOALS OVER TIME

AI is a fast-moving field, so your goals will likely evolve as the technology advances. New tools, breakthroughs, and even challenges will arise, forcing you to adjust your strategies. Staying adaptable and maintaining a growth mindset is critical to ensuring that AI remains aligned with your long-term ambitions.

For instance, if you've started using AI for content generation, you may later expand that into using AI for data analytics or strategy development. As you grow more comfortable with the technology, you'll find new ways to apply it to different areas of your life. The adaptability that AI demands also encourages a sense of openness to learning new things, keeping you agile and future ready.

Moreover, it's worth considering how the ethical implications of AI might affect your goals. For example, how can you ensure your use of AI tools is fair, transparent, and responsible? Aligning your personal values with the ethical use of AI is a step toward long-term success.

Action step: Schedule regular "goal audits" every three to six months. Revisit your initial goals and assess whether AI is helping you move closer to them. Adjust your strategy as needed to stay on track and in line with evolving AI capabilities.

Navigating the Emotional Landscape of AI: Embrace Possibility, Avoid Overwhelm

As with any significant technological shift, the rise of AI can bring emotional challenges alongside opportunities. It's easy to feel overwhelmed by the fast pace of change or to worry about job displacement, skill obsolescence, or ethical concerns. However, by focusing on how AI can enhance your own goals, you can turn these fears into opportunities for growth and development.

Rather than resisting the change AI brings, embrace it as a chance to evolve both personally and professionally. AI should be seen as an empowering tool, not as a replacement for human talent or creativity. The key is to continually look for ways in which AI can *supplement* your abilities rather than *supplant* them.

Surrounding yourself with a supportive community can also help ease some of the anxiety that comes with this new era. Join AI-focused forums, attend webinars, or collaborate with like-minded individuals who are also seeking to integrate AI into their lives.

Action step: Make a list of three ways you can turn your AI-related concerns into learning opportunities. For example, if you're worried about job displacement, consider learning about the new job roles AI is creating.

CONCLUSION: YOUR END GAME

When you align AI with your personal ambitions, you gain more than just a cutting-edge tool—you gain the ability to turn your unique goals into reality. AI provides the chance to elevate your skills, enhance your creativity, and broaden your horizons. The key is to stay focused on what truly matters to you, and to use AI as a vehicle to help you get there. Stay curious, keep learning, and let AI work in harmony with your dreams.

Chapter 18

SELF-EVOLVING

The AI Revolution
Adapting to the Ever-Changing AI Landscape

As the AI revolution reshapes industries, daily routines, and even creative practices, the key to thriving in this new age lies in constant evolution. The ability to self-evolve means continuously learning, adapting, and integrating new AI technologies into your personal and professional life. This chapter explores how to embrace change, nurture a growth mindset, and remain agile in the ever-evolving world of AI.

THE NEED FOR CONTINUOUS LEARNING

AI is moving at lightning speed, and what was groundbreaking technology yesterday may already be obsolete tomorrow. Staying competitive or relevant in an AI-driven world requires ongoing education. Whether you're a business leader, an artist, or just someone looking to harness AI for personal growth, it's crucial to keep learning.

Continuous learning doesn't have to be overwhelming. It can involve small, digestible pieces of information incorporated into your daily routine. For instance, subscribing to AI-related newsletters or podcasts can provide bite-sized knowledge that keeps you up to date. Online courses and communities, such as Coursera, Udemy, and even YouTube tutorials, can offer deeper learning opportunities.

Moreover, AI itself can be used to aid your learning process. Tools like *Duolingo* (for learning languages) or *ChatGPT* (for answering questions and explaining concepts) are built with AI-powered engines that personalize learning paths based on your pace and skill level. The use of AI in learning enables you to consume content in a way that fits your style and speed, offering personalized feedback and recommendations to fuel your growth.

Action step: Dedicate at least 30 minutes a day to learning something new about AI. This could be reading an article, watching a tutorial, or exploring an AI-powered learning platform.

BUILDING A GROWTH MINDSET

In the world of AI, flexibility and adaptability are paramount. The technologies we rely on today may look entirely different in just a years. Embracing a growth mindset—one that sees challenges as opportunities to learn—is essential for keeping up with AI's rapid evolution.

A growth mindset helps you see AI not as a threat to your job or creativity but as a tool that can help you improve and expand your potential. It encourages you to approach new AI developments with curiosity, rather than fear, and to view setbacks as opportunities for improvement. For instance, if a new AI tool initially feels difficult to use, instead of giving up, lean into the learning process. It may take time to get comfortable, but once you master it, you'll unlock new capabilities you didn't even know existed.

Building this mindset involves challenging your inner narrative. Instead of thinking, "I'm not tech-savvy enough to use AI," reframe your thoughts to, "I can learn to use this technology step by step." Adopting this approach will not only help you adapt to AI but also foster resilience in the face of change.

Action step: The next time you encounter a challenging AI tool, commit to mastering it rather than walking away. Document your learning process and note how overcoming the challenge expands your abilities.

PERSONALIZATION AND SELF-IMPROVEMENT: AI AS A TAILORED COACH

AI offers something that was nearly impossible to achieve in the past—highly personalized coaching. AI systems can now provide feedback that is specific to your personal or professional goals, making self-evolution more targeted and achievable.

For instance, AI fitness platforms like *MyFitnessPal* or *Peloton* tailor workouts to your fitness level and adjust as you improve, giving you real-time feedback on your progress. If you're focused on career growth, platforms like *LinkedIn Learning* recommend skills based on industry trends and your current expertise.

Creative professionals can benefit from AI too. Writers can use tools like *Grammarly* or *Sudowrite* to improve their work with instant feedback. Musicians can rely on AI platforms such as *AIVA* to suggest harmonies, melodies, or even complete tracks. These tools act as personalized coaches, helping you evolve and refine your craft with precision.

AI not only automates tasks; it provides insightful suggestions for improvement, empowering you to constantly evolve without a mentor or traditional support system.

Action step: Identify an AI tool that can act as your personal coach in one area of your life, whether it's fitness, creativity, or career. Start using it regularly to track your improvement.

AI-DRIVEN FEEDBACK LOOPS: MEASURING PROGRESS

To effectively evolve, you need to measure your progress. AI allows for seamless feedback loops that let you know exactly where you stand and where you need to improve. These feedback loops can come from diverse areas, whether in your work, your personal life, or your creative endeavors.

For example, AI-powered analytics in marketing can show you exactly how your campaigns are performing, pinpointing what's working and what isn't. Artists and designers can use tools like *Behance* or *Dribbble* to track audience engagement and adjust their creative direction. For personal improvement, wearable tech like *Fitbit* provides detailed insights into your health metrics—allowing you to tweak your routine for better outcomes.

The key benefit of AI-driven feedback is that it's data-backed, making it more reliable and actionable. Unlike traditional feedback, which may be subjective or delayed, AI delivers instant, objective insights, enabling faster course corrections and improvements.

Action step: Identify an AI tool that can provide real-time feedback in an area you're looking to improve. Make a habit of reviewing your performance data regularly and adjusting your actions based on these insights.

EMBRACING CHANGE: THE NEW NORMAL

Self-evolving is not just about improving; it's about embracing constant change. AI is a field that's evolving almost daily and staying stagnant means falling behind. As AI continues to transform industries, roles, and the way we interact with the world, adapting to this new normal is critical.

The good news is that AI itself can help you stay ahead of the curve. AI-driven news platforms, recommendation engines, and personalized content feeds are great ways to stay informed.

Platforms like *Feedly* can curate AI-specific news and updates, while *Twitter (X)* provides real-time conversations and emerging trends in the AI space. AI tools can also analyze your habits and suggest new skills you should learn to stay relevant in your field.

Moreover, many AI platforms allow you to set personal alerts for any significant changes in your industry. You can be notified of emerging trends, new technologies, or skills that are becoming increasingly relevant in the job market. By using AI to stay ahead, you'll always be ready to adapt when change inevitably comes.

Action step: Set up a personalized news feed or alert system that keeps you updated on AI trends and emerging technologies in your field. Make it part of your daily routine to check in on the latest advancements.

CONCLUSION: GROW WITH THE TIMES

In a world where AI is rapidly changing the way we live and work, self-evolution is not an option—it's a necessity. By committing to continuous learning, developing a growth mindset, and utilizing AI-driven feedback loops, you can ensure that you're constantly evolving to meet the demands of the future. AI provides the tools; it's up to you to stay agile, embrace change, and evolve alongside it.

Chapter 19

TURNING DREAMS INTO REALITY WITH AI

The AI Revolution
If You Can Think it, You Can Do It

We all have dreams—visions for our future, goals we aspire to, and creative projects we hope to bring to life. In the age of AI, turning these dreams into reality is not only possible but can be accelerated. By leveraging AI tools, individuals and organizations can harness the power of data, automation, and intelligent insights to breathe life into even the most ambitious goals. This chapter delves into how AI can help bridge the gap between ideas and execution, empowering anyone to turn their dreams into achievable outcomes.

DEFINING YOUR DREAM: CLARITY IS KEY

Every successful journey starts with a clear destination in mind. Before AI can assist you in realizing your dream, you must clearly define what that dream is. Are you hoping to start a new business? Perhaps you wish to become an artist, launch a tech startup, or need a 3D graphic for your prototype presentation. Defining your dream in concrete terms gives you a roadmap to follow.

Start by breaking down your dream into actionable steps. For instance, if your dream is to build an online business, the steps might include identifying your niche, understanding your audience, developing a website, and creating a marketing plan. Once you have a clear framework, AI can come into play to streamline and enhance each of these steps.

Action step: Write down your dream in as much detail as possible. Break it down into five key actions that will move you closer to achieving it. For each step, consider how AI could assist or automate part of the process.

USING AI FOR INSPIRATION: EXPANDING THE REALM OF POSSIBILITY

One of the most powerful ways AI can help turn dreams into reality is by expanding what you think is possible. AI is a source of creativity and innovation, often offering insights or ideas that you may not have thought of on your own. For example, AI-powered platforms like *OpenAI's DALL·E* allow artists to generate images based on textual descriptions, offering them new avenues for creative expression. Archaeologist can use tools like *ChatGPT* to brainstorm or even find out how the pyramids in Egypt collate with the stars.

AI can also enhance idea generation in more practical fields. For entrepreneurs, AI-driven market research tools can help identify underserved markets or predict future industry trends. Aspiring inventors can use AI to prototype new products by analyzing large datasets and suggesting optimizations. The possibilities are nearly endless once you start harnessing AI to amplify your creative process.

Action step: Use an AI tool to help brainstorm around your dream. Whether it's generating visual inspiration or drafting business ideas, let AI push the boundaries of what you think is achievable.

AUTOMATING MUNDANE TASKS: FREEING TIME FOR WHAT MATTERS MOST

Turning dreams into reality requires time, effort, and persistence. Many of us become bogged down by the repetitive, time-consuming tasks that inevitably come with any significant project. This is where AI excels: it can automate the mundane so that you can focus on the parts of your dream that truly require your creativity and expertise.

Consider how AI can streamline your workflow. For business owners, AI tools like *QuickBooks* can handle accounting tasks, while marketing automation platforms such as *MailChimp* can manage email campaigns. Creators can leverage AI for post-production work, whether it's editing videos with *Invideo* or retouching photos using AI-enhanced design tools. By automating these peripheral tasks, AI frees you to dedicate more time and energy to the high-value aspects of your dream.

Action step: Identify at least three routine tasks that AI could automate in your daily or weekly work. Explore AI tools that are suited to handle these tasks and start integrating them into your workflow.

AI AS A PARTNER IN PROBLEM SOLVING

Every big dream will encounter roadblocks along the way. Whether it's technical difficulties, financial challenges, or creative blocks, setbacks are inevitable. However, AI can be a valuable problem-solving partner. With AI-driven insights, you can approach challenges with data-backed solutions.

For instance, if you're an entrepreneur trying to attract customers, AI-powered analytics platforms like *Google Analytics* can provide insights into where your traffic is coming from and what content is most engaging. AI can also predict potential risks, allowing you to proactively avoid them. In creative fields, AI tools like *Adobe Sensei* can help streamline workflows, identify technical errors, and even suggest design improvements.

When it comes to personal challenges, AI-powered coaching apps such as *BetterUp* or *Woebot* can help you overcome mental blocks or emotional hurdles by offering personalized, evidence-based guidance.

Action step: Next time you face a challenge in pursuing your dream, explore how AI can provide insights, solutions, or new approaches. Use AI as a problem-solving partner to enhance your decision-making.

VISUALIZING YOUR DREAM: AI AS A CREATIVE PARTNER

Before any dream can be realized, it must first be clearly visualized. In the past, this step required significant time and resources, but AI has now made this process faster, more accessible, and more sophisticated. AI tools for artists, designers, and creators—like *MidJourney*, *RunwayML*, or *DALL·E*—can generate detailed images, concepts, or mock-ups based on a few descriptive prompts. These tools allow creators to visualize their ideas and iterate rapidly without needing to be an expert in design or illustration.

This accessibility opens the doors for dreamers who may not have the technical skills to bring their ideas to life in a traditional sense. An artist can now imagine an entire world, a character, or a narrative, and the AI will assist in generating visuals that closely match their vision. In the same way, entrepreneurs can use AI to model new products, architectural concepts, or even entire business plans.

Action step: Start by using an AI creative tool like *DALL·E* or *Ideogram*'s AI-powered features to generate a visual representation of your dream project. Use it as a source of inspiration to refine your vision further.

STREAMLINING THE PATH:
AI FOR PROJECT MANAGEMENT AND EXECUTION

Once you have a clear vision of your dream, the next challenge is execution. Many ideas remain just that—ideas—because people feel overwhelmed by the steps involved in turning them into reality. AI-powered project management tools, like *Trello*, *Monday.com*, or *Asana*, provide a solution by helping you break down your dream into actionable steps.

These tools use AI to prioritize tasks, predict deadlines, and provide insights into how you can complete your project more efficiently.

Beyond task management, AI also helps with logistics, budgeting, and resource allocation. For example, platforms like *GanttPRO* and *ClickUp* use machine learning to predict how long tasks will take based on previous patterns, helping you allocate your time and resources better. They also offer insights into how changes in one area of the project might impact the overall outcome, keeping your dream on track.

Action step: Break your project into smaller milestones using AI-driven project management tools. Use these tools to establish timelines, assign tasks, and track your progress, ensuring each step brings you closer to your ultimate goal.

FUELING CREATIVITY: AI AS A MUSE

One of the most powerful ways AI can assist in turning dreams into reality is by serving as a creative catalyst. AI algorithms can spark new ideas, suggest improvements, and even help you think outside the box in ways that human creativity alone may not achieve. For example, musicians can use AI platforms like *Amper Music* to generate melodies, harmonies, or background tracks that fuel new compositions. Writers can collaborate with AI text generators like *ChatGPT* to help refine plots, generate dialogue, or overcome creative blocks.

Creativity is no longer limited to an artist's personal capacity or inspiration. AI is a tool for iteration, allowing you to explore different versions of your work before settling on the perfect idea. This collaborative process enhances your capacity to produce innovative, high-quality work.

Action step: Use an AI creativity tool to generate new ideas or augment your current project. Play with different iterations or ask the AI to help you expand your original concept in unexpected ways.

COLLABORATING WITH AI: BRINGING TEAMS TOGETHER

Big dreams often require teamwork, and AI can be instrumental in fostering collaboration. AI-powered platforms like *Slack* and *Microsoft Teams* offer integrated collaboration tools that use AI to improve communication and teamwork. Features like real-time translation, automated meeting summaries, and intelligent task distribution ensure that everyone is on the same page and working efficiently toward a common goal.

AI collaboration tools also facilitate remote teamwork, which is essential in today's globalized workforce. Whether your team members are in the same room or on different continents, AI helps ensure that every voice is heard, and every contribution is valued.

Action step: Integrate AI-powered collaboration tools into your project management to enhance team coordination. Use features like automatic updates, reminders, and intelligent scheduling to streamline team communication and decision-making.

OVERCOMING OBSTACLES:
AI PROBLEM-SOLVING AND DECISION-MAKING

Every dream faces obstacles, but AI can serve as a powerful problem-solving tool. Whether you encounter logistical issues, creative roadblocks, or technical challenges, AI-powered systems can help you navigate through them. For example, if your project involves programming, tools like *GitHub Copilot* can assist with code generation and debugging. If you face financial challenges, AI budgeting tools like *Mint* or *YNAB* (You Need a Budget) can offer suggestions on how to cut costs or better allocate your resources.

AI excels at pattern recognition, making it an invaluable tool for decision-making. AI can analyze past trends, market conditions, or project data to offer insights that guide your next move. Whether it's deciding how to allocate resources or which strategy to pursue, AI can provide the data-backed rationale for your decisions.

Action step: When faced with a challenge, consult an AI-powered problem-solving tool relevant to your field. Use the insights it provides to inform your decisions and keep your project moving forward.

LAUNCHING YOUR DREAM: AI IN MARKETING AND OUTREACH

When it comes time to bring your dream to the world, AI can play a crucial role in amplifying your message. AI-powered marketing tools like *HubSpot*, *Hootsuite*, and *Google Ads* help optimize your marketing campaigns by analyzing audience behavior, predicting engagement, and recommending the best times and platforms to share your message.

Social media algorithms, fueled by AI, can help your project gain visibility by targeting the right audiences with personalized content. AI can also analyze user interactions to refine your marketing strategy and improve engagement. This makes your outreach efforts more effective and ensures your dream reaches its intended audience.

Action step: Use an AI-driven marketing platform to design and execute your outreach campaign. Let AI guide your decisions on when to post, who to target, and what content will generate the highest engagement.

INNOVATION THROUGH AI: REIMAGINING POSSIBILITIES

AI doesn't just help make existing ideas real—it enables entirely new forms of innovation. With AI tools like *GPT-4* or *OpenAI Codex*, innovators can automate complex tasks, build predictive models, or even develop entirely new applications. By leveraging AI's capacity for learning and pattern recognition, individuals can create products and services that wouldn't have been possible before, often with minimal technical expertise. Think of AI as a co-pilot, offering new avenues for experimentation, whether you're a creator, a business leader, or a startup founder.

For example, *AI-driven prototyping tools* can help bring hardware ideas to life. Platforms like *Figma* can use machine learning to automate elements of design, significantly cutting down development time while allowing you to focus on core innovations.

Action step: Tap into AI tools that offer code generation, market research, or customer interaction insights. Experiment with platforms that provide generative AI, like *Codex* or *Alethea AI*, to explore beyond what seems immediately possible.

OVERCOMING THE "IMPOSSIBLE" WITH AI: TACKLING BIG CHALLENGES

AI is an ideal tool for addressing what might have once seemed insurmountable challenges. Think of global issues like climate change, food security, and health, which often require solutions at scale and involve massive amounts of data. AI's capacity for data analysis, pattern recognition, and predictive analytics can break down these complex problems into solvable components.

Take the example of climate modeling: AI systems, like those used by organizations such as *DeepMind* or *IBM Watson*, can analyze vast amounts of climate data and predict future trends more accurately than human teams. On a smaller scale, individuals can tap into these datasets to craft local solutions, making their contributions impactful.

The same applies to personal dreams: You might be working on a sustainability project or a social initiative that feels beyond your grasp. By leveraging AI tools—whether for data analysis, crowdsourcing solutions, or resource management—you can address large-scale challenges while keeping a strategic, efficient approach.

Action step: Identify AI tools relevant to the challenges you face. Whether it's tackling a social issue or scaling a startup idea, explore AI-driven platforms for strategy and long-term planning, like *Zindi* for environmental data analysis or *C3.ai* for enterprise solutions.

THE EMOTIONAL INTELLIGENCE OF AI: ENHANCING PERSONAL GROWTH

While AI excels at technical tasks, it's also becoming adept at understanding human emotions. Emotional AI, like tools used in *customer service chatbots* or *virtual therapists*, reads tone, sentiment, and even body language. These systems can assist in improving emotional intelligence (EQ), which is crucial for personal and professional success. AI coaches like *Replika* or *Woebot* help individuals manage stress, set personal goals, and track emotional health.

For dreamers, this form of AI offers support by providing feedback on interpersonal relationships, helping with conflict resolution, and even fostering empathy. These systems can serve as mirrors, offering insights into your communication style or identifying emotional patterns that could hinder your growth.

Action step: Explore emotional intelligence AI platforms like *Woebot* or *Cloverleaf* to gain insights into your interactions, feelings, and growth potential. Use them to manage stress, improve communication, and stay emotionally balanced as you work toward realizing your dreams.

FROM CONCEPT TO MARKET: USING AI TO BUILD AND SCALE BUSINESSES

If your dream involves starting a business, AI can help you move from concept to market quickly. With AI-driven market analysis tools, such as *CB Insights* or *Gartner*, you can predict trends, evaluate competition, and understand your target audience more accurately. AI doesn't just help during the ideation phase but becomes indispensable in execution and scaling.

Once you're ready to launch, AI can automate customer support, manage supply chains, and optimize pricing models. For example, platforms like *Zendesk* integrate AI for customer service, while *Clearbit* uses AI for marketing insights. These tools allow you to focus on strategic decisions while the AI handles logistics.

Scaling can also be streamlined using AI-driven insights from sales platforms like *Salesforce Einstein* or *HubSpot*. By analyzing customer behavior, AI can suggest upselling opportunities or indicate which markets to target next, ensuring you grow efficiently.

Action step: Leverage AI-powered business tools like *Salesforce* or *HubSpot* to take your startup from local to global. Let AI help you automate marketing, sales, and customer service, while you focus on refining your product and strategy.

AI AND PERSONAL FULFILLMENT: A BALANCED APPROACH

Realizing your dreams isn't just about work or achievement; it's also about finding balance and fulfillment. AI can assist here too, by promoting healthier lifestyles and well-being through fitness tracking, mental health monitoring, and personalized wellness plans.

Wearables like *Fitbit* or AI-driven apps like *MyFitnessPal* offer tailored advice for staying physically healthy, while AI-based platforms like *Calm* or *Headspace* provide mental health resources. These tools go beyond generic recommendations, learning from your behaviors to offer personalized insights that can help you achieve well-rounded fulfillment as you chase your ambitions.

Action step: Utilize AI-driven health and wellness apps to maintain physical and emotional balance. These tools will support you in staying energized and mentally prepared for the challenges ahead.

SCALING UP WITH AI:
TAKING YOUR DREAM TO THE NEXT LEVEL

Once you've made significant progress toward achieving your dream, the next step is often to scale up. Scaling can mean different things depending on the dream—maybe it's growing a business, expanding a personal project into a global initiative, or even taking your art or writing to a broader audience. AI can play a crucial role in scaling by automating processes, offering predictive insights, and optimizing performance at a larger scale.

For entrepreneurs, AI-driven customer service platforms like *Zendesk* or chatbots like *Intercom* can handle thousands of customer queries at once, freeing up your team to focus on higher-level tasks. In creative industries, AI tools like *Canva Pro* or *Final Cut Pro X* allow you to mass-produce high-quality content in a fraction of the time it would take to do manually. Additionally, AI can analyze market trends and consumer behaviors, offering personalized recommendations for growth strategies.

Whether you're looking to scale up your personal brand, business, or creative project, AI provides the scalability that would otherwise require significant human and financial resources.

Action step: Consider how you want to scale your dream. Identify one area where AI can help you reach a larger audience, increase efficiency, or expand your impact. Implement one AI-driven strategy to start scaling up.

STAYING GROUNDED IN YOUR VISION:
AVOIDING THE AI OVERWHELM

While AI offers many advantages in turning dreams into reality, it's easy to become overwhelmed by the sheer amount of technology available. It's important to remember that AI is a tool, not a replacement for your creativity, vision, or passion. Staying grounded in your unique strengths will ensure that AI enhances your dream rather than diluting it.

As you integrate AI into your pursuit, make sure you remain the driving force behind your dream. AI can handle the data, the automation, and even some of the creativity, but only you can provide the vision, the heart, and the purpose behind the project. Use AI to lighten the load but never let it take over your dream.

Action step: Take a moment to reflect on your dream. Ensure that while you're using AI as a tool, you're staying connected to the core vision. Revisit your goals regularly to keep yourself on track.

CONCLUSION: DREAMS COME TRUE

With AI, turning your dreams into reality is more achievable than ever. By leveraging AI as a tool for inspiration, automation, problem-solving, and scaling, you can expedite the process of achieving your ambitions. However, the power of AI is most effectively harnessed when it aligns with your vision. Remember, AI is not the driver—you are. As you embark on this journey, stay true to your unique purpose, and let AI amplify your efforts in meaningful ways.

Chapter 20

AI ACCEPTANCE

The AI Revolution
*Navigating the Emotional and Practical Challenges
of AI Adoption*

As the AI revolution progresses, many of us find ourselves grappling with a mixture of excitement, apprehension, and confusion. The speed at which AI is integrated into our lives often feels dizzying, sparking questions not only about how we adapt practically but also how we emotionally process these changes.

Acceptance is a critical phase in this journey toward understanding and effectively working with AI. It allows us to embrace change, mitigate fear, and open ourselves to the possibilities that AI can offer. This chapter will explore both the emotional and practical dimensions of acceptance, helping readers develop a balanced perspective on integrating AI into their daily lives.

THE EMOTIONAL JOURNEY TO ACCEPTANCE

One of the biggest barriers to embracing AI is the fear of the unknown. We hear stories about jobs being lost to machines, the rise of autonomous systems, and the perceived dehumanization of work. These narratives can invoke fear and skepticism, clouding our ability to fully see the benefits AI can bring.

Understanding Fear and Resistance: Resistance to AI often stems from fear—fear of change, fear of job displacement, or fear of losing a sense of purpose.

It's essential to recognize these emotions as valid, but not let them dictate our actions. By acknowledging fear, we create space for curiosity and exploration, allowing us to slowly accept AI as a tool for growth rather than a threat.

The Anxiety of Replacement: Many people feel anxious about AI replacing jobs, but it's important to remember that technology has historically created new opportunities even as it disrupted old ones. For example, the automation of agricultural labor during the Industrial Revolution freed up time for education and new industries.

Losing Control:
Another common fear is that as AI becomes more powerful, we may lose control over our lives and work environments. Acceptance starts with realizing that AI is designed to augment human decision-making, not replace it. We can remain at the helm, using AI as a powerful assistant.

Fostering Emotional Resilience:
Reframe the Narrative: Instead of viewing AI as a threat, consider it as a partner in progress. AI can help us eliminate tedious tasks, freeing up time for creative and meaningful work.

Mindfulness and Adaptation: Techniques such as mindfulness can help us manage anxiety about AI. By staying present and focusing on what we can control, we become better at handling the rapid changes happening around us.
Practical Acceptance: Integrating AI into Everyday Life

Practical acceptance means actively seeking out ways to incorporate AI into our lives. We've already seen AI pop up in the form of smart home devices, virtual assistants, and personalized recommendations, but the potential goes far beyond that.
The question becomes, how do we integrate AI in a way that complements our lifestyle and professional goals?

Start Small:
Acceptance doesn't mean diving headfirst into complex AI systems. Begin by incorporating small AI tools into your everyday routine. This could be something as simple as using a virtual assistant like Siri or Google Assistant to manage your daily tasks or experimenting with AI-powered note-taking apps like Otter.ai to transcribe meetings and save time.

Build Trust Gradually:
Trusting AI is a gradual process. Initially, it's natural to feel skeptical or hesitant about allowing AI to handle certain tasks. Start by giving AI small responsibilities, then gradually increase its involvement as you feel more comfortable. For example, allow your AI assistant to manage your schedule before letting it assist with more complex tasks like drafting emails or making financial decisions.

Balancing Human Judgment and AI Automation:
It's crucial to remember that AI doesn't need to replace human judgment—it should complement it. AI excels at processing vast amounts of data, identifying patterns, and automating repetitive tasks. However, human intuition, empathy, and creativity remain irreplaceable. By striking a balance between AI automation and human oversight, we ensure that AI enhances our decision-making rather than diminishing it.

The Role of Acceptance in Professional Growth

As AI reshapes industries, professionals in every field must accept that staying relevant requires continuous adaptation. Acceptance in the workplace means actively seeking out opportunities to integrate AI tools, learning to work alongside them, and understanding the new skills needed to thrive in an AI-enhanced environment.

Lifelong Learning in the Workplace:
Incorporating AI into the workplace demands a commitment to lifelong learning.

Employees should seek out AI-related training and professional development opportunities to stay ahead of the curve. Employers, in turn, should offer support by creating learning environments where employees can safely experiment with new AI tools.

Redefining Roles and Responsibilities:
As AI takes over repetitive tasks, human roles will evolve to focus more on strategic thinking, creativity, and interpersonal communication. Professionals who accept these changes and focus on developing uniquely human skills will find themselves more valuable than ever in an AI-driven economy.

Collaboration Between Humans and AI:
We must learn to view AI as a collaborator rather than a competitor. For instance, AI can handle data analysis, while humans can interpret the results and make informed, empathetic decisions. In healthcare, for example, AI can assist doctors by diagnosing diseases from medical images, but the human doctor is needed to explain the diagnosis to the patient, provide emotional support, and make nuanced treatment decisions.

THE SOCIETAL SHIFT TOWARD AI ACCEPTANCE

AI adoption isn't just an individual challenge—it's a societal one. As AI technologies become increasingly embedded in our infrastructures, from transportation to healthcare, we must collectively work toward acceptance at a community and governmental level.

Policy and Regulation:
Governments play a crucial role in guiding AI adoption by establishing regulations that ensure the ethical use of AI. Acceptance on a societal level involves recognizing the need for oversight, transparency, and accountability in the deployment of AI systems. We must support policies that protect privacy, ensure data security, and prevent biased or unethical AI applications.

Creating a Culture of Innovation:
A culture of innovation is key to societal acceptance of AI.
Communities, industries, and educational institutions should
foster environments where experimentation with AI is encouraged.
Whether it's hackathons, startup incubators, or collaborative
research initiatives, we must work to create spaces where people
feel empowered to explore AI's potential.

CONCLUSION: THE PATH TOWARD ACCEPTANCE

Acceptance of AI is a journey, one that requires emotional
resilience, practical exploration, and a societal shift in mindset.
By confronting our fears, embracing lifelong learning, and actively
integrating AI into our daily lives, we can navigate the AI revolution
with confidence.

Rather than resisting change, we should lean into it, seeing AI as
a tool that can help us grow, both personally and professionally.
AI has the potential to improve our quality of life, elevate human
creativity, and open doors to opportunities we never thought
possible. But to unlock these benefits, we must first accept AI into
our lives.

Chapter 21

YOUR PERSONAL ASSISTANT

The AI Revolution
Embracing AI as a Partner, Not a Threat

As the rapid rise of artificial intelligence continues to reshape industries and redefine the future of work, one of the biggest challenges people face isn't about skillsets or learning new technologies. It's about accepting AI's role in our lives. Fear, uncertainty, and even skepticism surround AI, and these emotions are often fueled by misconceptions about the technology's potential dangers, as well as its ability to replace jobs or threaten personal privacy. However, much like any transformative tool in human history—from the printing press to the internet—acceptance and understanding are key to benefiting from what AI has to offer.

OVERCOMING FEAR AND MISCONCEPTION

For many, AI conjures up images of dystopian futures where machines dominate, replace, or control humans. While these fears are largely born from science fiction, they still influence real-life anxiety about AI's growing presence. It's natural to feel cautious about unfamiliar technology, especially when headlines sometimes emphasize risks more than opportunities. However, understanding the nuances of AI's capabilities, ethical guidelines, and its limitations is crucial to dispelling these misconceptions.

UNDERSTANDING THE ROLE OF AI

AI, at its core, is a tool. Its primary goal is to augment human capability, not replace it. The distinction between artificial intelligence and human intelligence needs to be made clear: AI is exceptional at performing repetitive, data-driven tasks, but it lacks the creativity, empathy, and nuanced thinking that define human intelligence. By accepting AI as an assistant rather than an adversary, we can refocus our energies on how to best collaborate with it.

Demystifying AI Fears:

*Job Displacement vs. Job Transformation: While AI will automate certain tasks, it will also create new job opportunities. The key lies in reskilling and upskilling workers to fit into roles that AI will complement rather than eliminate. The history of industrial revolutions shows that while some jobs vanish, new and often more rewarding professions emerge.

*AI as a Decision-Making Partner: AI can assist in decision-making by analyzing vast datasets, but it doesn't "decide" on its own. It provides insights, leaving the final call in human hands, ensuring control remains with people.

*Data and Privacy: AI systems today are subject to stringent data protection regulations. Ethical AI development includes privacy safeguards that protect individuals' rights, ensuring AI operates within a framework of trust.

VIEWING AI AS A COLLABORATIVE FORCE

Once we've moved past our fears, it becomes easier to see AI as a partner rather than a competitor. Much like computers and the internet revolutionized the way we work, communicate, and live, AI has the potential to act as a transformative ally in almost every aspect of life.

AI IN THE WORKPLACE

Incorporating AI into professional environments leads to increased efficiency and productivity. Automating repetitive tasks allows workers to focus on more strategic, creative, and value-driven initiatives. For example, AI in healthcare assists doctors by analyzing medical scans faster and more accurately, freeing up time for direct patient care. AI in legal practices can sift through legal documents, highlighting relevant cases in seconds, empowering lawyers to focus on strategy and case-building.

THE HUMAN-AI TEAM DYNAMIC

The future of work isn't about humans or AI but about how we work together. By allowing AI to handle mundane, data-heavy tasks, individuals are freed to tap into their uniquely human skills—creativity, problem-solving, and emotional intelligence. This collaborative approach not only improves outcomes but also provides personal satisfaction, as workers can spend more time on fulfilling, high-value activities.

In industries like education, AI can assist teachers in creating personalized learning plans while allowing educators to focus on mentoring and hands-on guidance. In marketing, AI can analyze consumer data, providing insights that allow human marketers to craft more effective, creative campaigns.

EMOTIONAL ACCEPTANCE:
COMING TO TERMS WITH AI'S UBIQUITY

It's one thing to intellectually understand AI's benefits, but emotional acceptance can take longer. AI is here to stay, and instead of resisting its growth, learning to embrace its presence can lead to more positive experiences.

THE EVOLUTION OF ATTITUDES TOWARD TECHNOLOGY

When the internet was first introduced, many people resisted it, concerned about how it would change communication, jobs, and social interactions. Yet today, the internet is not only accepted but embraced as a necessity for daily life. Similarly, attitudes toward AI are evolving. Those who once feared automation now recognize its potential to create value, free up time, and enhance capabilities. This emotional evolution takes time, but the more we interact with AI, the more it becomes part of our normal landscape.

EMBRACING LIFELONG LEARNING

Part of accepting AI is accepting that learning never stops. We live in an age where continuous upskilling is crucial. The idea that formal education ends after school is outdated. Embracing AI means embracing a mindset of lifelong learning, curiosity, and adaptability. Online courses, mentorships, workshops, and real-world experimentation allow us to grow alongside AI, continually refining our skills and remaining relevant in this new era.

ETHICS AND RESPONSIBILITY IN AI ADOPTION

AI's power also comes with significant responsibility. As more companies, governments, and individuals adopt AI technologies, it's crucial that we think critically about the ethical implications of its use. Acceptance of AI isn't just about acknowledging its potential benefits—it's also about committing to using AI in ways that are fair, transparent, and responsible.

ETHICAL AI DEVELOPMENT

Developers and companies must commit to creating AI that is unbiased, inclusive, and designed to serve all of society. Avoiding algorithmic biases and ensuring diverse data sets are critical steps toward achieving fairness in AI outputs. Likewise, transparency in AI decision-making allows users to trust the systems they interact with.

AI GOVERNANCE

Governments and policymakers must regulate AI usage to prevent harmful applications, ensuring that AI is used for societal good rather than for invasive surveillance, misinformation, or exploitation. Public input into how AI systems are deployed in healthcare, criminal justice, and finance can help ensure accountability. It's important for society to engage in conversations about the rules and standards that should govern AI's role in our lives.

PRACTICAL STRATEGIES FOR EMBRACING AI

Here are some actionable ways to emotionally and practically embrace AI's role in your life:

*Start Small: Begin by incorporating AI into simple daily tasks. Whether using an AI-powered assistant like Siri or Alexa to manage your schedule or experimenting with AI-driven design tools like Canva, familiarizing yourself with AI in small doses makes it feel less intimidating.

*Engage with AI Communities: Join AI-focused groups, attend webinars, or participate in forums like Reddit's r/MachineLearning. Engaging with a community of like-minded individuals can provide support, knowledge sharing, and a sense of camaraderie in navigating AI's complexities.
*Recognize Its Limitations: AI is not a panacea. By acknowledging where AI excels and where it falls short, you can set realistic expectations. This will help prevent frustration and disappointment when AI doesn't perform as expected and reinforce its role as a tool, not a magic solution.

CONCLUSION: AI, A NEW ASSISTANT ON YOUR TEAM

By reframing AI as a partner in growth and development, we open ourselves to its vast potential. With acceptance comes empowerment, and by embracing this transformative technology, we can actively shape a future where humans and AI collaborate for the greater good.

Chapter 22

A NEW HOPE

The AI Revolution
How AI Can Improve the Future for Humanity

In every technological revolution, there are moments of
uncertainty, but there are also moments of immense hope. AI, with
its vast potential, stands on the cusp of such a moment. Beyond the
anxiety and apprehension, lies a future filled with possibilities for
improving human life, addressing long-standing global challenges,
and creating a brighter tomorrow.

This chapter explores the hope AI offers, not only in technological
advancement but also in its capacity to transform society, health,
the environment, and our collective future.

AI AND THE FUTURE OF HEALTHCARE

Healthcare is one of the sectors where AI holds the most promise.
The technology has already begun revolutionizing the way we
diagnose, treat, and prevent disease, but the possibilities extend far
beyond what we've seen today.

Personalized Medicine: AI's ability to process vast amounts of
genetic, medical, and lifestyle data allows for truly personalized
healthcare solutions. We are moving toward a future where
treatments are specifically tailored to an individual's unique
biology. Diseases like cancer, diabetes, and heart conditions can be
better managed through data-driven predictions about how specific
treatments will work for each patient.

Predictive Analytics: AI can predict potential health issues before they fully manifest. For example, wearable devices that monitor heart rates, blood pressure, and glucose levels are coupled with AI models that can predict heart attacks or strokes days or even weeks before they occur. This means earlier interventions and potentially life-saving actions.

Medical Imaging and Diagnosis: AI is being used to analyze medical scans more accurately than the human eye alone can. For instance, AI systems can detect early signs of cancer in mammograms, identify fractures, and even recognize rare diseases from X-rays and MRIs. These technologies could improve diagnostic accuracy and dramatically reduce waiting times for patients.

Addressing Global Health Disparities: AI also presents an opportunity to bring healthcare to underserved populations. In rural areas or developing nations where healthcare infrastructure may be lacking, AI-powered telemedicine systems can provide diagnostic services through a smartphone. AI-based chatbots, for example, can help triage patients, offering guidance on whether they need to seek immediate medical attention or whether their symptoms can be treated at home.

With AI, we envision a world where top-tier medical care is accessible to everyone, not just those in well-resourced areas.

AI'S ROLE IN ENVIRONMENTAL SUSTAINABILITY

AI has already begun to play a crucial role in combating climate change and promoting environmental sustainability. By using data to monitor ecosystems, optimize energy consumption, and even predict natural disasters, AI could be a game-changer for the environment.

Optimizing Energy Use:
AI can be applied to reduce energy consumption across industries. For example, smart grids that use AI can optimize electricity distribution, balancing supply and demand more efficiently. In homes and offices, AI-powered thermostats and lighting systems automatically adjust to save energy, minimizing waste while maintaining comfort.

Sustainable Agriculture:
AI can also help address issues like food security and water scarcity by optimizing agriculture. AI-driven sensors in fields can monitor soil health and water needs in real-time, recommending the exact amount of water and fertilizer needed for optimal plant growth, reducing waste. Drones equipped with AI systems can monitor crop health, detect diseases early, and even help in precision farming techniques that maximize yields while minimizing environmental damage.

Climate Prediction and Disaster Response:
AI's ability to process vast amounts of climate data can help scientists model and predict environmental changes more accurately. AI-powered models are already being used to predict natural disasters such as hurricanes, floods, and wildfires. More accurate predictions allow for better preparedness and quicker response times, ultimately saving lives and reducing the economic impact of these disasters.

Conservation Efforts:
AI-driven drones and remote sensing technologies can help monitor endangered species and illegal deforestation activities, protecting biodiversity. AI can analyze patterns in poaching activity, track animal populations, and help in the overall preservation of ecosystems critical to the health of our planet.

AI AND THE FUTURE OF WORK

One of the most significant questions surrounding AI is its impact on jobs. While concerns about job loss are valid, it's essential to recognize that AI can create new opportunities for employment, many of which we have not yet envisioned.

Redefining the Workplace:
Rather than eliminating jobs, AI will likely change the nature of work itself. The focus will shift from manual, repetitive tasks to more creative, strategic roles. Workers will need to harness emotional intelligence, creativity, and problem-solving skills—areas where AI still cannot replace humans.

The Rise of New Professions:
Just as the Industrial Revolution created entirely new industries and job roles, AI will lead to the rise of new professions. Fields such as AI ethics, AI auditing, and AI-human interaction design will emerge, offering roles that bridge the gap between human capabilities and AI's potential.

Empowering the Workforce:
AI can empower workers to perform their jobs more efficiently. For instance, AI can handle administrative tasks like scheduling meetings, analyzing data, or managing inventories, freeing employees to focus on higher-value activities. Workers will need to learn how to collaborate with AI systems, leveraging their abilities to increase productivity and innovation.

Upskilling and Lifelong Learning:
AI's rapid integration into the workforce will require a strong emphasis on education and upskilling. Workers will need to embrace continuous learning, acquiring new skills to stay relevant. Governments and industries will need to collaborate on creating educational programs that equip the workforce with the skills needed to thrive in an AI-driven economy.

AI'S CONTRIBUTION TO GLOBAL PROBLEM SOLVING

AI has the potential to tackle some of the world's most pressing problems in ways that were previously unimaginable. Whether it's addressing poverty, improving education, or solving the global hunger crisis, AI offers new avenues for innovative solutions.

AI in Education:
AI-powered educational platforms can revolutionize the way we learn, offering personalized learning experiences that adapt to each student's pace and style. These systems can also help bridge gaps in access to education by providing low-cost, scalable learning resources to underserved communities.

Tackling Global Poverty:
AI can be used to analyze economic data, pinpointing regions where poverty is most severe and helping organizations distribute resources more efficiently. AI-driven microfinance platforms, for instance, are providing small loans to entrepreneurs in developing countries, empowering them to start businesses and improve their livelihoods.

Global Hunger and Resource Allocation:
AI's predictive capabilities can be used to optimize resource allocation in disaster-prone or poverty-stricken regions. By analyzing satellite imagery and agricultural data, AI systems can predict food shortages and guide humanitarian efforts more effectively. In regions suffering from food insecurity, AI can also suggest crop patterns, resource distribution, and even delivery logistics.

CONCLUSION: EMBRACING AI FOR A HOPEFUL FUTURE

While AI presents challenges, it also offers us unprecedented hope. By harnessing AI's capabilities, we can address some of the most significant problems humanity faces. Healthcare can become more equitable, the environment can be preserved, and industries can evolve to meet the needs of tomorrow.

The future is not one of AI versus humans—it is a future of AI and humans working together to create a better, more sustainable, and more hopeful world.

Chapter 23

RESOURCES FOR NAVIGATING THE AI REVOLUTION

The AI Revolution
Get Going! You're Already Late

As we've explored throughout this book, AI presents incredible opportunities, but navigating the AI revolution requires the right tools and knowledge. In this chapter, we will explore some of the key resources that will help you harness AI's power in different areas of life, work, and personal growth. Whether you're just beginning your journey into AI or are already familiar with its possibilities, the resources outlined here will help you stay informed and engaged.

LEARNING PLATFORMS AND EDUCATIONAL TOOLS

One of the first steps in mastering AI is acquiring foundational knowledge. Fortunately, there are numerous online platforms where you can learn at your own pace, with or without a technical background.

Massive Open Online Courses (MOOCs):
MOOCs are great for learners who want structured lessons delivered by top-tier universities and companies. Some platforms offer free courses, while others have premium options with certificates.

Coursera: Coursera partners with universities like Stanford, Princeton, and companies like Google to offer courses on everything from AI basics to specialized topics like machine learning and deep learning.

Courses such as Andrew Ng's "Machine Learning" are highly popular and accessible even to those without prior experience.

edX: Similarly, edX offers courses from Harvard, MIT, and other leading institutions. Their "AI for Everyone" and "Introduction to Python for AI" are excellent starting points for newcomers.

Udacity: Known for its nanodegree programs, Udacity has AI-focused courses like "Artificial Intelligence for Robotics," developed in collaboration with companies like Nvidia. These programs often include project-based learning, giving students hands-on experience.

YouTube Channels and Tutorials:
If formal courses aren't your style, YouTube is an excellent free resource for AI learning. Channels like *3Blue1Brown* offer visual explanations of complex mathematical concepts used in AI, while *Sentdex* focuses on Python programming and AI tutorials. These can help break down difficult topics into digestible, everyday lessons.

AI TOOLS AND SOFTWARE FOR NON-CODERS

You don't need a degree in computer science to leverage AI in your life or business. Many AI-powered tools are designed to be user-friendly, enabling people from various fields to harness AI's power without deep technical knowledge.

AI Art and Creativity Tools:
Artists, writers, and creators can now use AI to enhance their craft. Some of these tools are as simple as uploading an image or text and allowing the AI to generate creative outcomes based on your input.

RunwayML: This tool allows artists and designers to create AI-generated art, videos, and animations without needing to code. With Runway, you can bring your creative visions to life by collaborating with AI-powered models.

OpenAI's DALL-E and GPT: OpenAI's DALL-E can generate unique images from simple text descriptions, while GPT-3 is a language model that can write coherent text, helping writers with blog posts, storytelling, and even brainstorming new ideas. Both tools are easy to use and powerful for creators who want to explore the intersection of AI and art.

Chatbots and Conversational AI Tools: Services like *Landbot* and *Tars* make it easy for businesses to create AI-driven chatbots for customer service without requiring any coding knowledge. These tools can enhance customer engagement by automating responses to frequently asked questions and guiding users through their services or products.

AI COMMUNITIES AND NETWORKING OPPORTUNITIES

Connecting with others who share your interest in AI is an important way to stay updated on the latest advancements, find job opportunities, or collaborate on projects. AI is a rapidly evolving field, and staying connected to the community can help you keep pace with new developments.

Meetups and Online Communities:
Joining a local or online AI group can help you build relationships with others, share knowledge, and stay motivated in your learning journey.

Meetup.com: This platform hosts numerous AI-focused meetups in cities around the world. From coding bootcamps to ethical AI discussions, you can find a group that suits your needs and interests.

Kaggle: If you're interested in data science and machine learning, Kaggle is a community and competition platform where you can work on AI problems and challenges alongside others. Beginners can use it as a learning tool, while experts may engage in advanced AI projects to showcase their skills.

Reddit (r/MachineLearning, r/Artificial): These subreddits are vibrant communities where AI enthusiasts, professionals, and beginners gather to discuss AI news, share research, and seek advice on AI projects.

PROFESSIONAL NETWORKS

For those looking to advance their career in AI, joining professional organizations can provide valuable networking opportunities and access to conferences and publications.

The Association for the Advancement of Artificial Intelligence (AAAI): This global organization promotes research in AI and provides educational resources, conferences, and a prestigious AI journal.

Women in AI: An organization dedicated to increasing diversity in the AI field, offering mentorship, networking events, and training programs for women pursuing careers in AI.

STAYING INFORMED ON AI ETHICS AND GOVERNANCE

As AI continues to play a more prominent role in our lives, understanding the ethical implications and governance surrounding AI is crucial. There are several organizations and resources dedicated to fostering responsible AI use, which can help you stay informed and engaged in the ethical debates that shape AI's future.

Ethical AI Research Centers:
These centers are committed to exploring the ethical, legal, and societal impacts of AI, providing insights that help shape public policy and corporate AI strategies.

The AI Now Institute: A research institute at New York University, AI Now focuses on the social implications of AI technologies. They publish reports on issues like surveillance, bias in AI algorithms, and the future of work, all of which are critical for understanding how to build fair and equitable AI systems.

The Partnership on AI: This multi-stakeholder organization is committed to ensuring that AI technologies are developed in ways that benefit people and society. With members ranging from academic institutions to major tech companies, the Partnership on AI offers reports, best practices, and guidelines for ethical AI development.

AI Governance Platforms:
Staying informed on AI governance is essential for professionals working in policy, business, or leadership roles.

AI Policy Labs: An organization that helps governments and businesses create and adopt policies for the responsible use of AI. It offers resources and case studies on AI ethics, governance, and regulation.

World Economic Forum's Global AI Council: This council offers a broad global perspective on how different countries and industries are approaching AI governance. Their publications are an excellent resource for understanding how policies surrounding AI are evolving.

BOOKS AND PUBLICATIONS

For those who prefer in-depth learning through books, there are several highly regarded publications on AI, covering both technical and non-technical topics. Some books provide a comprehensive overview of AI's evolution, while others dive deep into its impact on society, business, and ethics.

Technical Books:
- "Artificial Intelligence: A Modern Approach" by Stuart Russell and Peter Norvig: This is considered the definitive textbook on AI. While highly technical, it's an essential read for anyone looking to deeply understand AI algorithms, machine learning, and computational intelligence.

Deep Learning" by Ian Goodfellow, Yoshua Bengio, and Aaron Courville: This book offers a comprehensive introduction to deep learning and is used widely in academic and industry settings. It's perfect for those who want to take a deeper dive into AI's subfield of machine learning.

Non-Technical Books:
AI Superpowers" by Kai-Fu Lee: This book offers insight into how AI is transforming industries across the globe, particularly in China and the U.S. Lee also provides an accessible discussion on how AI will shape the future of work.

Life 3.0: Being Human in the Age of Artificial Intelligence" by Max Tegmark: Tegmark explores the philosophical and societal questions raised by AI, from existential risks to human consciousness. A thought-provoking read for those interested in the broader implications of AI on humanity.

CONCLUSION: BUILDING YOUR AI ARSENAL

As AI continues to evolve, so must our approaches to learning, adapting, and leveraging its power. With the right resources, you can stay ahead of the curve, ready to embrace the transformative power of AI and thrive in a world where it is increasingly woven into the fabric of our lives. Whether you're a beginner or a seasoned professional, there are endless opportunities to dive deeper, grow your understanding, and play a part in shaping the AI revolution.

A LETTER TO THE FUTURE

Congratulations! You've just taken a remarkable journey through the ever-evolving landscape of Artificial Intelligence. What you've read isn't just a guide, it's a call to action. A revolution is here, and you're standing at its forefront.

The AI revolution isn't a distant concept—it's now. It's transforming how we live, work, and dream. But here's the truth: it's not about machines replacing humanity; it's about humans expanding their potential. You are equipped with the knowledge and tools to not just survive but *thrive* in this new era.

You may have started this book feeling overwhelmed by AI's complexity, perhaps even fearful. But look at you now: informed, inspired, and ready to take on challenges you may not have thought possible. AI is your ally in shaping a better future—whether it's enriching your personal life, amplifying your creativity, or propelling your career to new heights. You're ready to embrace the opportunities it offers while maintaining the human touch that makes you unique.

Remember, the possibilities are endless, but your curiosity, your adaptability, and your openness to change are the keys to unlocking them. Take what you've learned here and apply it to your life, your work, your family. Share it with others. Challenge yourself to grow with AI, not against it.

The future belongs to those who see AI not as a threat, but as a bridge to a better, more connected, more innovative world. And you—yes, you—are part of that world.

Thank you for walking this path with me. The revolution has just begun, and I can't wait to see how you'll thrive in it.

With optimism and excitement for what's to come.

Dr. Lawrence Capri

GLOSSARY

1. Artificial Intelligence (AI): Machines mimicking cognitive functions such as learning and problem-solving.

2. Machine Learning (ML): A branch of AI where systems learn from data and improve over time.

3. Neural Networks: Systems modeled after the human brain, consisting of layers of nodes (neurons) that process data.

4. Deep Learning: A complex type of ML with multiple layers of neural networks.

5. Natural Language Processing (NLP): AI that enables machines to understand and generate human language.

6. Supervised Learning: A learning method where the AI is trained on labeled data to make predictions.

7. Unsupervised Learning: AI learns from unlabeled data by identifying patterns on its own.

8. Reinforcement Learning: Learning where AI takes actions in an environment and is rewarded or penalized based on outcomes.

9. Training Data: Data used to train an AI model.

10. Algorithm: A set of rules or calculations followed by a machine to solve a problem.

11. Big Data: Extremely large data sets that are analyzed computationally to reveal patterns and trends.

12. Data Mining: The process of extracting useful information from large datasets.

13. Bias in AI: When an AI system produces prejudiced or unequal outcomes due to flawed training data or algorithm design.

14. API (Application Programming Interface): A set of functions allowing applications to communicate with each other, often used for integrating AI.

15. Generative AI: AI capable of creating content, such as text, images, or music, from scratch.

16. Inference: The process of applying a trained AI model to new data to make predictions or decisions.

17. Chatbot: An AI application designed to engage in conversation with humans.

18. Robotics: A field of technology related to the design and use of robots, often incorporating AI for autonomous decision-making.

19. Turing Test: A test to determine if a machine's behavior is indistinguishable from that of a human.

20. Deepfake: Synthetic media where a person in an existing image or video is replaced with someone else's likeness, often using AI.

21. Cognitive Computing: AI systems that aim to simulate human thought processes.

22. Data Labeling: The process of annotating data to be used for training supervised learning models.

23. Feature Engineering: The process of selecting, modifying, or creating input features for a machine learning model.

24. Overfitting: When a machine learning model performs well on training data but fails to generalize to new, unseen data.

25. Underfitting: When a model is too simple to capture the underlying patterns in the data, leading to poor performance.

26. Autonomous Vehicles: Vehicles capable of navigating and operating without human input, often powered by AI.

27. Cloud Computing: Storing and accessing data and programs over the internet instead of on local computers, facilitating scalable AI.

28. Explainability: The ability to explain how AI models reach their conclusions, crucial for trust and ethical use.

29. Edge AI: AI computation performed on devices like smartphones or IoT devices, closer to where data is generated.

30. Transfer Learning: A technique where a model developed for one task is reused as a starting point for another task.

31. Model Training: The process of teaching an AI model to recognize patterns by feeding it data.

32. Computer Vision: A field of AI that enables machines to interpret and understand visual information from the world.

33. Pattern Recognition: The ability of AI to detect regularities in data, forming the basis of many AI applications.

34. Recommender System: An AI system that suggests products or content based on user behavior.

35. Tokenization: The process of breaking text into smaller units, such as words or subwords, for processing by NLP models.

36. Anomaly Detection: The identification of rare events or observations that differ significantly from the majority of the data.

37. Sentiment Analysis: The use of AI to interpret and classify emotions expressed in text data.

38. AI Ethics: The principles guiding the responsible development and deployment of AI, including fairness, privacy, and accountability.

39. Autonomous Agents: AI systems capable of independent decision-making, such as self-driving cars or delivery robots.

40. Clustering: A type of unsupervised learning where data is grouped into clusters based on similarity.

41. Convolutional Neural Networks (CNNs): A type of deep learning neural network used primarily in image recognition tasks.

42. Generative Adversarial Networks (GANs): AI systems consisting of two neural networks that compete to improve each other's performance, often used to create realistic images.

43. Swarm Intelligence: AI systems modeled on the behavior of decentralized, self-organized systems, such as ant colonies.

44. Quantum Computing: Emerging computational technology that uses quantum-mechanical phenomena to perform calculations, potentially revolutionizing AI processing.

45. Backpropagation: A training algorithm used for neural networks where errors are propagated backward through the network to improve performance.

46. Multimodal AI: AI models that integrate and process data from multiple sources, such as text, images, and audio.

47. Federated Learning: A decentralized approach to training machine learning models, where data is kept locally, and only model updates are shared.

48. Ethical AI: The pursuit of building AI systems that adhere to moral and ethical standards to avoid harm and promote fairness.

49. Human-in-the-Loop: Systems that combine AI decision-making with human oversight, ensuring AI outputs align with human values.

50. Digital Twins: Virtual models of real-world entities or systems, often powered by AI, used to simulate, predict, and optimize outcomes.

Family & Education-Related Terms

51. AI Tutors: Virtual systems that provide personalized learning experiences for students.

52. Personalized Learning: Educational methods where AI tailors lessons based on individual students' needs.

53. EdTech (Educational Technology): The use of AI and digital tools to enhance learning environments.

54. Smart Classrooms: AI-enabled classrooms with interactive technologies for enhanced learning experiences.

55. Educational Chatbots: AI-based assistants that help students with questions or problems in real time.

56. VR Learning: The use of virtual reality environments for immersive education experiences.

57. Emotional AI: AI systems designed to understand and respond to human emotions, aiding in emotional learning and interaction.

58. Adaptive Learning: Systems that modify educational material in real-time to match a learner's performance.

59. AI in Special Education: AI tools that assist students with disabilities in accessing and understanding educational content.

60. Digital Homework Assistance: AI platforms that provide support to students in completing their homework assignments.

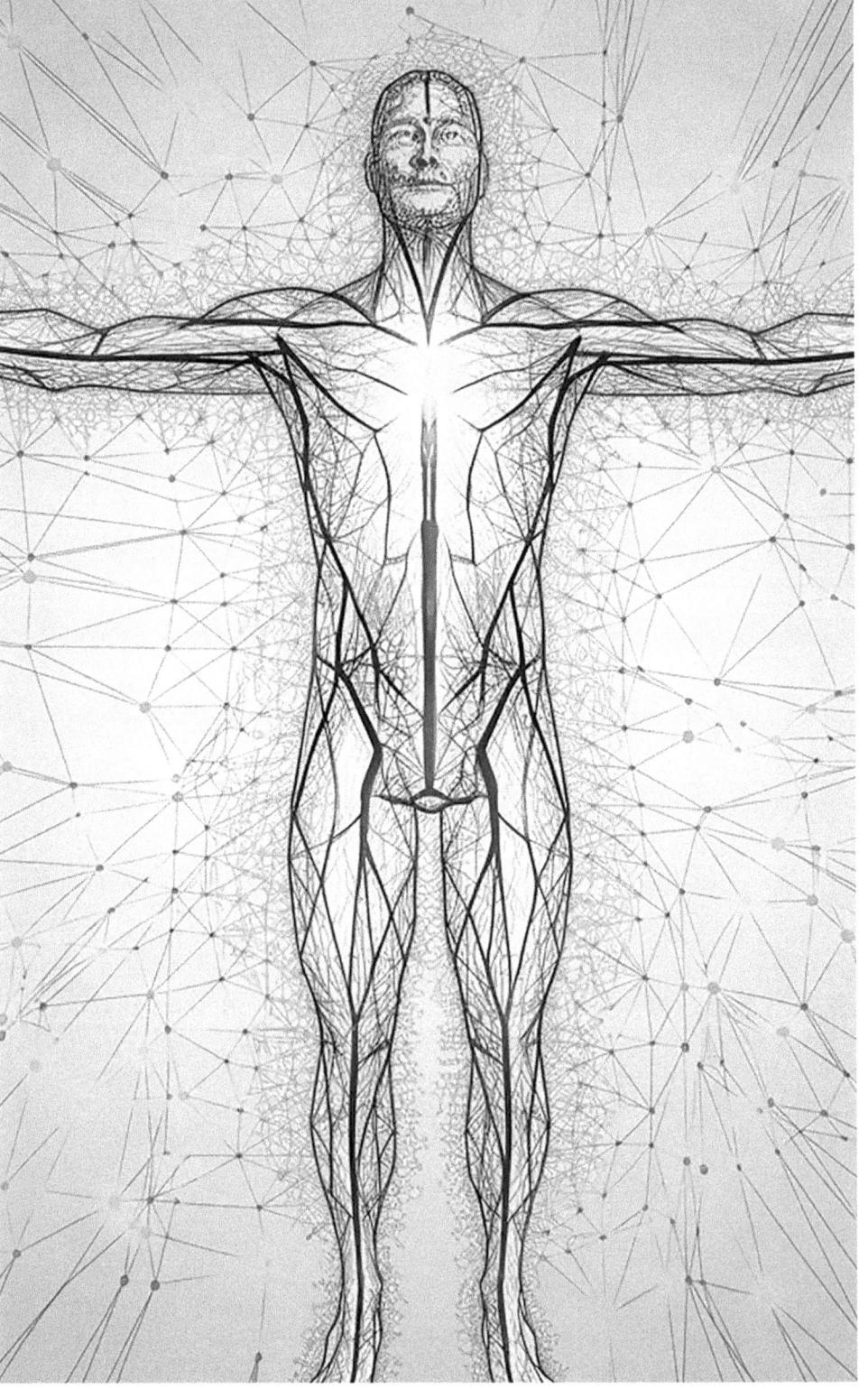

THE REVOLUTION HAS BEGUN

WWW.AIMYEVERYTHING.COM

WWW.YOUTUBE.COM@DR.LAWRENCECAPRI

www.ingramcontent.com/pod-product-compliance
Lightning Source LLC
Chambersburg PA
CBHW051619120626
46551CB00014B/1868